Encyclopedic Dictionary of
Pipeline Integrity

Guy Desjardins • Reena Sahney

CLARION
TECHNICAL PUBLISHERS

Encyclopedic Dictionary of
Pipeline Integrity

Printed in the United States of America.

ISBN 0-9717945-6-1

Preface

User guide: This book is intended as a reference guide for the broad range of topics that fall under the umbrella of "pipeline integrity." Thus, it has been written with an eye to practicality, and with an attempt at simplification of often complex concepts, so the reader may get a quick overview of a topic or term and understand perhaps what questions next to ask. As such, we have taken liberties in presenting some definitions that may not be strictly accurate or complete to a technical specialist in the subject but that are sufficient to provide the reader a quick and practical grasp of key points. We hope the result is a helpful and easy-to-use reference for pipeline integrity practitioners.

Acknowledgments

Contributions: We thank David Horsley, Dr. Chris Kedge, Dr. Alan Murray, Mo Mohitpour, and John Tiratsoo for their time and effort in providing input as well as reviewing the work. Further, the figures were patiently constructed based on (often) rough concepts by the team at ParanoidFish Graphic Design. We are grateful also to the authors and publishers who permitted us to use or adapt material from publications and course materials that were primary sources, in particular Bill Bruce, Dr. Thomas Bubenik, Dr. Phil Hopkins, and Dr. Roger King.

Inspiration: The inspiration for this book came from the *Encyclopedic Dictionary of Applied Geophysics* by Robert Sheriff. His work reminded us of the usefulness of having a quick reference guide that allows us to grasp the practicalities of a topic in a relatively short period of time – providing us a powerful tool for determining what needs further study.

Publisher: The infinite patience and guidance of our publisher provided the freedom, latitude, and means to explore and realize our vision for the book.

Permissions Acknowledgment

We are most grateful also to the authors and publishers who permitted us to use or adapt material from publications and course materials that were primary sources, in particular Bill Bruce, Dr. Thomas Bubenik, Dr. Phil Hopkins, and Dr. Roger King, and the following organizations:

American Petroleum Institute

American Society of Mechanical Engineers

ASM International

Crane Co.

DAMA International

Det Norske Veritas

Elsevier

John Wiley & Sons

Macaw Engineering

NACE International

Penspen Group

Pipeline Operators Forum

Prentice-Hall

About the Authors

Guy Desjardins has more than 27 years' experience in the oil and gas industry and 14 years with pipelines. He is managing director of Desjardins Integrity Ltd, a consulting firm offering services in the analysis of in-line-inspection data, risk and reliability of various types of pipeline systems. Guy holds a Bachelor's degree in Geophysics from the University of Alberta in Edmonton. He is a professional member of the Association of Professional Engineers, Geologists and Geophysicists of Alberta (APEGGA) and has been active with a number of industry organizations including NACE International and API.

Reena Sahney brings 20 years of experience in the energy sector to a consulting practice that encompasses pipeline engineering education, pipeline integrity, project management and the assessment and quantification of risk associated with operating capital intensive assets. She is a mechanical engineer by training, with a Masters Degree from Carleton University (focused on non-destructive testing) as well as an MBA from Queen's University. Reena has worked in the North American Energy Sector as well as the UK with organizations such as TransCanada, Deloitte and the Southern Alberta Institute of Technology (SAIT), before transitioning to consulting with Jiva Group, Inc. Industry participation has always been a significant part of her career – Reena chaired the development of the original NACE Recommended Practice on Inline Inspection – and she continues with participation in the International Pipeline Conference and, at this writing, serving as a member of the ASME Pipeline Systems Division's Executive Committee.

Table of Contents

Encyclopedic Dictionary of Pipeline Integrity 1

Reference Figures

Figure 1. Illustration of 100mV polarization criterion...1
Figure 2. 3t x 3t interaction rule applied to corrosion anomalies...............................2
Figure 3. A-scan display of ultrasonic testing results (1)..4
Figure 4. Activation polarization at the anode site..8
Figure 5. Illustration of the ALARP principle (4)..9
Figure 6. Angle-beam reflection from a crack. (1) 12
Figure 7. Schematic of mist flow. (10) 13
Figure 8. Axial, radial, and circumferential directions on a pipeline..........................18
Figure 9. B-scan display...20
Figure 10. Plot illustrating the difference between B31G and modified B31G equations...........22
Figure 11. Ball valve schematic (a) and operation (b). (200) 23
Figure 12. Bathtub failure curve..25
Figure 13. Beta distribution for range of α and β. The exponential distribution is where α=1. ..27
Figure 14. Schematic of bubble flow. (10) 30
Figure 15. Relevant parameters for calculation of Bulk modulus [adapted from (10)]. 31
Figure 16. Butterfly valve schematic (a) and operation (b) (200)..............................33
Figure 17. Cap undercut. (197)..36
Figure 18. Galvanic vs. impressed current cathodic protection systems. (195) 39
Figure 19. Swinging check valve schematic (a) and operation (b). (200)....................44
Figure 20. Chi-squared distribution for a variety of degrees of freedom....................45
Figure 21. Approximate timeframes for usage of various coating types. (181)...........49
Figure 22. Concentration polarization..56
Figure 23. Confidence interval. 57
Figure 24. ILI report girth-weld numbering after a cut-out repair. 67
Figure 25. Cylindrical coordinate system...67
Figure 26. The main elements of data management systems as identified by DAMA. (114) 69
Figure 27. Dead leg and potential liquid/debris collection points..............................70
Figure 28. Principle behind the operation of a dead-weight pressure tester. 70
Figure 29. Classification of corrosion anomalies (defects) (16)..................................72
Figure 30. Relationship between background noise signal strength and potential pipe anomalies. 75
Figure 31. Four-step direct assessment process..78
Figure 32. Drip leg used to collect liquids in a gas line. 81
Figure 33. Schematic of drop weight tear test. (198) 82
Figure 34. Anode and cathode in a galvanic cell (18). 87
Figure 35. Endurance limit on an S-N diagram. 89
Figure 36. Excess concavity in a weld. (197)..91
Figure 37. Excess convexity in a weld. 91
Figure 38. Excess penetration in a weld. 92
Figure 39. Exponential distribution for various lambda values. 93
Figure 40. Simplified external corrosion direct assessment process (3)......................94
Figure 41. Comparison of the Folias factor for B31G and modified B31G....................100

Figure 42. Fracture and gas decompression velocity as a function of pipeline pressure [adapted from (8), courtesy P. Hopkins]...101

Figure 43. Gamma distribution for various values of α and β.
The special case of α=1 is the exponential distribution...105

Figure 44. Gate valve schematic (a) and operation (b). (200) 107

Figure 45. Schematic indicating location of girth weld...110

Figure 46. Principle of globe valve operation. 111

Figure 47. Example of a grind repair [adapted from (8)]..114

Figure 48. Plot of PDF when using Gumbel distribution. 116

Figure 49. Illustration of the Hall effect used for MFL sensors. 117

Figure 50. Comparison of Brinnell, Vickers, and Rockwell C hardness tests [adapted from (195)].
119

Figure 51. Cross-section of a weld showing the heat-affected zone (HAZ). 120

Figure 52. Formation of hook crack from lamination in ERW procedure (195)...........................123

Figure 53. The process for installing a temporary bypass using hot tapping (9). 124

Figure 54. Example of hot tap assembly. (6) 125

Figure 55. Principle of hydrogen flux monitoring. 127

Figure 56. Lack of penetration in weld. (197)..138

Figure 57. Key steps in liquid penetrant inspection. 141

Figure 58. Methods of pipeline design. (9)..142

Figure 59. Lognormal distribution for various values of μ and σ. The special case of α = 1 is the exponential distribution. 143

Figure 60. Simplified schematic of magnetic flux leakage pig. 145

Figure 61. Summary of magnetic particle inspection process. (1) 146

Figure 62. Misalignment at a weld. 150

Figure 63. Misplaced cap on a weld. 151

Figure 64. Plot of PDF when using normal distribution..161

Figure 65. Use of o'clock position to indicate the orientation of an object on the pipeline 162

Figure 66. Outer diameter on a pipeline..166

Figure 67. Partially turbulent flow. (180)..168

Figure 68. Principle of Pearson survey [adapted from (195)]...170

Figure 69. Typical pig launcher/receiver facilities. 175

Figure 70. Schematic of various pipeline systems. 180

Figure 71. Principle of plug valve operation. (200)..183

Figure 72. Various dimensions used for calculating Poisson's ratio. (10) 185

Figure 73. Example of discrete probability distribution. 193

Figure 74. Example of continuous probability distribution. 193

Figure 75. "Tail" behavior for selected probability distributions used in pipeline integrity risk and reliability analysis..193

Figure 76. Probability of exceedance is calculated as the area under the probability-density function above the critical depth...195

Figure 77. Example of river bottom profile used for corrosion burst pressure calculations (6).207

Figure 78. Root pass on a weld. 208

Figure 79. RStreng® input requirements (6)...209

Figure 80. Input values for calculation of shear modulus. (10) 215

Figure 81. Slug flow. (10) 219

Figure 82. Stratified flow. (10) 225

Figure 83. Stress vs. shear stress (23)..226

Figure 84. Geometry (i.e., stress concentrator) impact on flow stress in pipe wall (25)...........227

Figure 85. Stress-strain curve for mild steel [adapted from (23)]. 229

Figure 86. Mechanism of sulfide stress cracking [adapted from (25)]. 230

Figure 87. Tenting phenomenon (10) 234

Figure 88. Single-phase laminar and turbulent flow in a pipeline. (10) 241

Figure 89. Transverse vs. longitudinal ultrasonic waves. (1)....................243

Figure 90. Various ultrasonic testing arrangements and corresponding signals (1)...............244

Figure 91. Example of in-line inspection tool velocity plot with speed excursion.....................247

Figure 92. Schematic of wavy stratified flow. (180) 251

Figure 93. Plot of PDF when using Weibull distribution. 252

Figure 94. Inputs for calculating of Young's modulus. (10)....................259

ReferenceTables

Table 1. Subparts of the US "Transportation of Hazardous Liquids by Pipeline"regulation (182)....................3

Table 2. Selected factors affecting method of pipeline abandonment (183) 5

Table 3. Definitions of active corrosion. Adapted from (3)....................8

Table 4. Common assessment criteria for corrosion (8) 17

Table 5. Comparison of B31G and modified B31G equation assumptions (8)....................22

Table 6. Major subtypes of ball valves (27) 24

Table 7. Beta distribution properties....................26

Table 8. Major types of biocides used in pipeline applications (18)....................27

Table 9. Composition specifications for carbon steels (195) 37

Table 10. Comparison of galvanic and impressed current cathodic protection systems (186)....................39

Table 11. Four main cathodic protection criteria (5) 40

Table 12. Major subtypes of check valves (27)....................44

Table 13. Chi-squared distribution properties....................45

Table 14. Class locations per CFR 192.5 (13) 47

Table 15. Selected factors affecting corrosion in the presence of carbon dioxide (195)....................48

Table 16. Selected considerations in selecting a pipeline coating (168)....................50

Table 17. Summary of coating types, strengths and weaknesses (168) 53

Table 18. Construction defects and related integrity concerns [adapted from (14)] 58

Table 19. Selected approaches to establishing corrosion growth rate 59

Table 20. Main corrosion mapping techniques [adapted from (15)]....................60

Table 21. Three stages of crack life (10) 63

Table 22. Various types of cracking in coating (15) 63

Table 23. Key components of unrefined crude oil (10) 65

Table 24. Typical defect classification categories for an MFL pig run (124)....................72

Table 25. Typical components of a dig sheet (124) 77

Table 26. NACE Standards associated with direct assessment (184)....................78

Table 27. Key considerations in designing a dual-diameter pig (124) 82

Table 28. Three key mechanisms in internal pipe wall erosion (10)....................90

Table 29. Exponential distribution properties 93

Table 30. Various single and multiphase flow regimes in pipelines (10) 98

Table 31. Impact of various parameters on fracture velocity (8)....................101

Table 32. Impact of various parameters on gas decompression velocity (8)....................102

Table 33. Properties of gamma distributions 105

Table 34. Indication of the composition of natural gas mixtures (8)....................106

Table 35. Major subtypes of gate valves (27)....................107

Table 36. Key skills required in geomatics engineering (138)....................109

Table 37. Common forms of geotechnical threats on a pipeline (26) 109

Table 38. Considerations when assessing gouges (8)....................112

Table 39. Mechanical properties of common pipe grades (20)....................113

Table 40. Gumbel distribution parameters....................116

Table 41. Common causes of hard spots in pipelines. (8) 118

Table 42. Equivalence of Brinnell, Vickers, and Rockwell C hardness tests [adapted from (195)] 119

Table 43. Key differences between standard and high-resolution corrosion pigs (124)............122

Table 44. Most common methods used to detect coating holidays (195) 123

Table 45. Conditions needed for hydrogen cracking (6)...125

Table 46. Major types of hydrogen damage on pipelines (8) 126

Table 47. Sources of hydrogen in pipelines (195)...126

Table 48. Effect of H_2S on humans [adapted from (11)] 128

Table 49. Most common inhibitor types for pipeline applications and the associated mechanism. (18) 131

Table 50. Key considerations in establishing an inspection interval for smart pigs.................133

Table 51. Lognormal distribution properties..142

Table 52. Tool depth grading for Standard-resolution MFL tools 144

Table 53. Common mill-related anomalies (8, 10, 195)..150

Table 54. Major types of risk mitigation..151

Table 55. Common examples of multiphase pipeline products (180) 153

Table 56. Pipeline integrity problems, and potential solutions, associated with common multiphase systems (180)...............................154

Table 57. Outer pipeline diameter versus NPS (9)...158

Table 58. Summary of pipeline NDT methods [adapted from (1)] 159

Table 59. Relative uses and merits of various NDT methods [adapted from (1)].....................160

Table 60. Normal distribution properties..161

Table 61. Typical conditions favorable to obligate anaerobes. (18) 162

Table 62. Key characteristics of various types of liquid penetrant (1)..........................170

Table 63. Key characteristics of various types of permafrost (26)................................171

Table 64. Major pig types and typical uses [adapted from (10)]..................................173

Table 65. Key elements of pig trap design (124) 175

Table 66. Key elements of a procedure used for pipeline pigging (124) 177

Table 67. Key elements of a "pipe tally" typically provided by in-line inspection vendors (124)..179

Table 68. Three main types of pipelines: distribution, gathering and transmission (8) 181

Table 69. Summary of common pipeline monitoring methods and their applications [adapted from (8)] 181

Table 70. Key elements of US *Pipeline Safety Act* (182) 182

Table 71. Key elements of a pre-(pig)run questionnaire (124)....................................189

Table 72. Advantages and disadvantages of pressure testing (10)..............................191

Table 73. Standard pressure testing requirements (ASME B31.8) (10) 191

Table 74. Common probability functions for pipeline reliability analysis (187) 194

Table 75. Key factors in determining probability of failure (196) 195

Table 76. Common repair methods (6) 203

Table 77. Typical reporting requirements for in-line inspection reports and associated excavations (124) 204

Table 78. Corrosivity dependence on soil resistivity (18) 204

Table 79. Qualitative risk matrix (31) 206

Table 80. Advantages and disadvantages of RStreng® technique (8).............................209

Table 81. SCC Density definitions (24)..212

Table 82. Comparison of common pipeline repair sleeve types (6) 218

Table 83. Key conditions/factors for the occurrence of SCC (25) 227

Table 84. Categorization of SCC (24)..227

Table 85. Key components of a SCADA system (10) 231

Table 86. Surface preparation standards most relevant for pipeline integrity related work (186) 232

Table 87. Nine major terrain types [adapted from (26)] 235

Table 88. Major controllable threats and mitigations for an existing pipeline [adapted from (8)] 238

Table 89. Examples of constraints that could result in a pipeline being considered "unpiggable" (124) 245

Table 90. Major valve types [adapted from (27) and (10)].......................................248

Table 91. Summary of water jetting categories based on pressure (186) 250

Table 92. Weibull distribution properties..252

Table 93. Common welding processes for pipeline applications [adapted from (6)] 253

Table 94. Common weld defects (14) 255

Table 95. Advantages and disadvantages of weld deposition repair (6) 256

NACE Standards...261

API and ASME Standards 274

References..290

0-9

1ˢᵗ party: Refers to the operator of the pipeline. The term is most commonly used with reference to safety incidents as part of clarifying the association(s) of the individual(s) involved in the incident.

2ⁿᵈ party: Refers to an authorized contractor of the pipeline operator. The term is most commonly used with reference to safety incidents as part of clarifying the association(s) of the individual(s) involved in the incident.

3ʳᵈ party: Refers to unauthorized parties, often the general public, involved with some aspect of the pipeline. The term is most commonly used with reference to safety incidents as part of clarifying the association(s) of the individual(s) involved in the incident.

100mV polarization criterion: This is one method used to assess whether adequate cathodic protection is being applied to a pipeline or other buried structure. The criteria states that the cathodic protection is adequate if, after the CP current is interrupted, there is a negative 100mV shift of the electrical potential readings. The potential shift is calculated by subtracting the measured steady-state off potential from the *instant off* potential (see Figure 1).

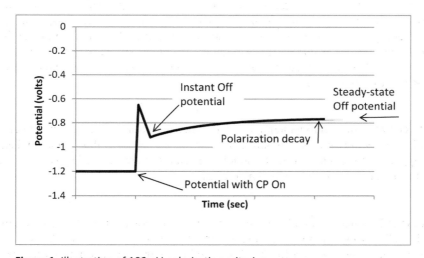

Figure 1. Illustration of 100mV polarization criterion.

3t×3t interaction rule: This term refers to the analysis and assessment of corrosion and the potential interaction of individual anomalies. The 3t×3t is an interaction rule used to group individual corrosion anomalies detected by an in-line inspection tool into *clusters*. See Figure 2. 3t x 3t interaction rule applied to corrosion anomalies. The rule states that by expanding the size of the anomalies in all directions by a distance 3 times the wall thickness of the pipe, if two or more expanded boxes overlap, then the anomalies interact. The collection of all interacting anomalies forms a cluster. At times, this method is often referred to as the 6t x 6t method since the actual distance between the anomalies is 6t.

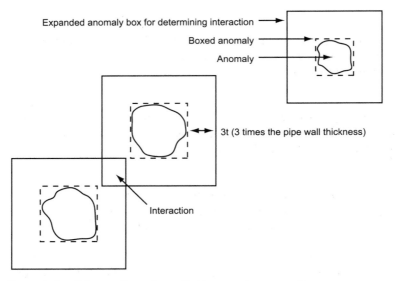

Figure 2. 3t x 3t interaction rule applied to corrosion anomalies.

49 CFR Part 192: American regulation entitled "Transportation of Natural and Other Gas by Pipeline: Minimum Federal Safety Standards." The scope of this regulation includes the minimum safety requirements for pipeline facilities and the transportation of gas including pipeline facilities and the transportation of gas within the limits of the outer continental shelf as that term is defined in the Outer Continental Shelf Lands Act (43 U.S.C. 1331). Gases regulated by this act include natural gas, flammable gas, and gas that is toxic or corrosive.

49 CFR Part 193: American regulation entitled "Transportation of Natural and Other Gas by Pipeline: Minimum Federal Safety Standards." The scope of this regulation includes a pipeline facility that is used for liquefying natural gas or synthetic gas or transferring, storing, or vaporizing liquefied natural gas.

49 CFR Part 195: American regulation entitled "Transportation of Hazardous Liquids by Pipeline." The scope of this regulation includes the minimum safety standards and reporting requirements for pipeline facilities used in the transportation of hazardous liquids or carbon dioxide. The regulation covers pipeline facilities in or affecting interstate or foreign commerce, including pipeline facilities on the Outer Continental Shelf. Hazardous liquids regulated by this act include petroleum, petroleum products/distillates, supercritical carbon dioxide, and anhydrous ammonia. Subparts of this regulation cover:

Table 1. Subparts of the US "Transportation of Hazardous Liquids by Pipeline" regulation (182)

Subpart	Topic
A	General
B	Annual, Accident, and Safety-Related Condition Reporting
C	Design Requirements
D	Construction
E	Pressure Testing
F	Operation and Maintenance
G	Qualification or Pipeline Personnel
H	Corrosion Control

6t interaction rule: This term is a way of referring to the 3t×3t interaction rule when describing the clustering of corrosion data from an internal inspection tool. See *3t×3t interaction rule*.

-850mV (polarization) criterion: This is one method used to assess whether adequate cathodic protection is being applied to a pipeline or other buried structure. It states that if the measured *instant off potential* (See Figure 1) of the pipeline is less than or equal to -850mV with respect to a copper-copper sulfate reference electrode, then the pipeline is adequately protected.

A

A-scan: One of the presentation formats used to display *ultrasonic testing* results where signal amplitude is plotted against time (as per Figure 3.). The time of the reflection (x-axis) is directly related to the thickness of the sample under investigation.

$$ToFL = 2 \times \frac{thickness}{v}$$

where   *ToFL*   = time of flight
             v   = velocity of the ultrasonic signal in the material
     thickness   = width of the material (i.e., distance from the transducer to the reflective surface)

The division by 2 is required because the wave makes a round-trip from the transmitter/receiver transducer to the refection and back to the transducer.

See also *B-scan* (plot of signal amplitude vs. one-dimensional view, such as length or width of the component) and *C-scan* (representation of apparent corrosion on a two-dimensional view of the component).

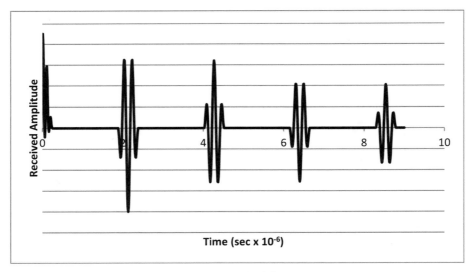

Figure 3. A-scan display of ultrasonic testing results (1).

abandonment: The permanent removal of a pipeline or other installation from service. In most jurisdictions abandonment can be done either by physical removal of the pipeline or by leaving the structure in-situ – having met the appropriate regulations and requirements. It is highly unlikely that either method alone will be used for the entire pipeline segment based on how conditions vary with geography and land use. Regardless of the method chosen, abandonment requires certain actions to be undertaken to ensure environmental and public safety of the abandoned installation. (2) The primary determinant in choosing in-situ abandonment or physical removal is existing as well as future land use; however, several factors should be considered – including those itemized in Table 2.

Table 2. Selected factors affecting method of pipeline abandonment (183)

Factor	Description
Future land use	If land is likely to remain undeveloped, then in-situ abandonment may be acceptable; however, the presence of population or other sensitive structures may drive the full removal of the pipeline.
Ground subsidence	If ground subsidence is not a concern, in-situ abandonment may be acceptable; however, it should be noted that the removal of the pipeline could actually increase the risk of subsidence if remediation is inadequate. Subsidence is a greater concern for large diameter pipe – should the pipe (eventually) collapse, it may provide a channel for the flow of water and increased soil erosion.
Soil/groundwater contamination	The long-term risk of soil and groundwater contamination may drive the decision towards full-scale removal of the pipeline.
Foreign crossings	The existence of a large number of foreign crossings (e.g., crossing of other pipelines and utilities) may favor the in-situ abandonment alternative.
Cost	Typically in-situ abandonment is a lower cost alternative; however, other factors or special considerations (e.g., potential for soil and groundwater contamination) may deem complete removal of the line a more cost effective approach.

above-ground installation: Any installation associated with a pipeline system that is not buried. This usually consists of mainline valves, compressor / pumping facilities and major interconnections.

above ground marker (AGM): This is a portable device placed on the ground above a pipeline, usually at a surveyed location, that detects and records the passage of an in-line inspection tool; an AGM may also be configured such that it transmits a signal that is detected and recorded by the tool. The AGM records the time at which the inspection tool passes its location. That location is used as a reference point in determining the location of other buried pipeline features.

abrasive blast cleaning: Also referred to as abrasive blasting, this is the cleaning and roughening of a surface produced by the high-velocity impact of an abrasive that is propelled by the discharge of pressurized fluid from a blast nozzle or by a mechanical device such as a centrifugal blasting wheel.

The abrasive material consists of small particles, such as sand, crushed chilled cast iron, crushed steel grit, aluminum oxide, silicon carbide, flint, garnet, or crushed slag, which is propelled at high velocity to impact a surface.

absolute risk: This term refers to the element of risk (associated with a specific pipeline section for example) in absolute terms – that is, in relation to other societal risks (e.g., fatality in a car accident). A quantified, absolute risk value is calculated as opposed to a *relative risk* ranking of pipeline segments. As a result, this approach allows an operator to manage risk in a manner that can be broadly understood and compared to other risks that individuals in society may be exposed to.

AC: See *alternating current.*

accelerant: This is a chemical substance that initiates or accelerates the rate of a chemical reaction. An accelerant's behavior is similar to that of a catalyst, with a key difference in that an accelerator may be consumed by the reaction process. In the context of pipeline integrity, accelerants are commonly used to increase the rate of polymerization or curing of a coating on a pipeline. The accelerant increases the rate at which cross-linking of the polymer occurs, or causes polymerization to occur at a lower temperature. Also used interchangeably with the terms activator and accelerator.

acceptable risk: A generic term which is usually defined based on the specific context of the risk assessment. That is, acceptable risk is a function of many different factors including (but not limited to): Who is exposed to the risk? Is the exposure voluntary or involuntary? What is the perception of risk? What is the level of risk aversion? In the context of pipeline integrity, the vast majority of discussions associated with acceptable risk are tied to probability of leak and rupture and establishing tolerable levels of these failure modes across the length of the pipeline system. There have been a number of attempts to standardize this term in a rigorous way – the most notable being the effort by the *Health and Safety Executive (UK)*. See *ALARP*.

accumulation horizon: See *soil.*

accuracy: The correctness of a measurement or the closeness of a measurement of a quantity to its true value. The difference between a measured value and the true value is the error. Since the size of the error is not generally known, the accuracy is stated as error bounds with a confidence level. For example the accuracy of the depth of corrosion might be stated as ±0.2 mm, 80% of the time. The stated accuracy has two parts: the tolerance and the confidence level (also called the certainty). In this example, the tolerance is 0.2 mm and the confidence level is 80%. The stated accuracy indicates that for 80% of the measurements, the absolute value of the error is less than 0.2 mm.

acid gas: A gas with a significant amounts of hydrogen sulfide (H_2S) or carbon dioxide (CO_2) and, as such, its presence presents an increased risk of internal corrosion. In a pipeline, acid is produced by a reaction with water.

In the case of hydrogen sulfide, hydrogen sulfide is a weak acid and one or two hydrogen ions are produced directly from the hydrogen sulfide.

$$H_2S \rightarrow H^+ + HS^- \rightarrow 2H^+ + S^{2-}$$

In the case of carbon dioxide, it reacts with the water to form carbonic acid. Like hydrogen sulfide, carbonic acid is a weak acid, which donates one or two hydrogen ions.

$$CO_2 + H_2O \rightarrow H_2CO_3 \rightarrow H^+ + HCO_3^- \rightarrow 2H^+ + CO_3^{2-}$$

acoustic emission corrosion monitoring: Acoustic emission is a method of monitoring corrosion by "listening" for (i.e., detecting the emission of acoustic) sound waves within the material. Some corrosion processes such as stress corrosion cracking emit sound waves because of the mechanical release of stresses. These sound waves can be detected and their presence indicates growth of the crack anomaly; however, the technique is somewhat limited due to relatively high background *noise* levels in the relevant range.

actionable anomaly: This is any anomaly that exceeds acceptable limits based on the analysis of the data for that anomaly and pipeline as defined by the API 1163 "In-line Inspection Systems Qualification Standard."

activation polarization: Also termed the *polarized potential* or *on potential*, this is a component of the total polarization at an electrode. Thus, the term is relevant for interpreting and applying cathodic protection criteria.

Specifically, The total electrochemical polarization is the difference between the electrical potential and the equilibrium potential. The polarization is given by the equation:

$$\eta_{tot} = E - E_{eq}$$

where $\quad \eta_{tot}$ = polarization

$\quad E_{eq}$ = equilibrium potential

$\quad E$ = potential across the surface of the metal

The total polarization, η_{tot}, is the sum of the potentials from three processes:

$$\eta_{tot} = \eta_{act} + \eta_{conc} + iR$$

where $\quad$ = activation potential

η_{conc} = concentration potential

iR = potential drop due to the flow of the current through the circuit

Activation polarization is a function of the energy required to drive the *half-cell* reaction, and is given by

$$\eta_a = \beta_a \log \frac{i_a}{i_0}$$

where $\quad i_a$ = current density at the anode

$\quad i_0$ = exchange current density

$\quad \beta_a$ = Tafel constant

The activation polarization is illustrated in Figure 4. Activation polarization at the anode site. The activation potential is the difference between the corrosion potential and the equilibrium potential as indicated by the arrow.

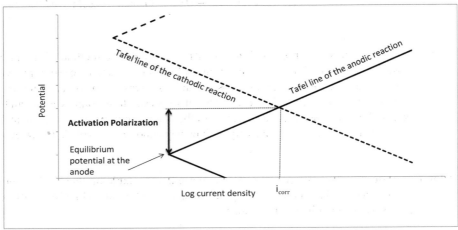

Figure 4. Activation polarization at the anode site.

active corrosion: Several definitions exist for this term; some more practical than others – as per Table 3.

Table 3. Definitions of active corrosion. Adapted from (3)

Nature of definition	Definition
OPS (Regulatory)	"….Corrosion which if left unattended could adversely affect the safe operation of the pipeline…."
Engineering	A State of electrochemical activity which will promote corrosion growth and continued wall loss
Practical	Corrosion that is not sufficiently controlled by cathodic protection

actuator: A powered *valve operator*. More specifically, actuators are a key component of the automation of valves as they physically operate (i.e., open and close) the valve based on an external signal. There are three main categories of actuators (and many variations within these basic categories, depending on the situation): they are typically either pneumatically (relying on pressurized gas), hydraulically (relying on pressurized liquids) or electrically powered.

ACVG: See *alternating current voltage gradient*.

aerial survey: Refers to the periodic visual inspection of the pipeline *right-of-way (ROW)* from the air to identify potential threats or pipeline integrity related concerns, such as encroachment, third-party mechanical damage, or unplanned release of the pipeline product. It is also used to survey, inspect and select ROW alternatives. A variety of sensors may also be deployed, such as photographic, video or even hydrocarbon detectors.

aerobic: A term used to indicate the presence of readily available oxygen – usually used to describe an environment, process, or life form. An aerobic environment is one where oxygen is readily available for chemical processes. An aerobic process is a process that requires the presence of oxygen. Aerobic life (such as aerobic bacteria) is life that requires oxygen to survive; various forms of aerobic life can or cannot tolerate a lack of oxygen. From a pipeline integrity point of view, whether an environment is aerobic or *anaerobic* can be important in understanding (and mitigating) *microbiologically influenced corrosion.*

AGA: See *American Gas Association.*

AGM: See *above ground marker.*

air drying: Refers to an external coating process where the applied wet coat is transformed to a dry coating film through the evaporation of solvent, or reaction with oxygen, i.e., simple exposure is sufficient and the addition of heat or curing agent is not required to complete the process.

air test: A pressure test conducted with air (as opposed to liquid or another fluid). See *pressure test.*

ALARP: This term, an acronym for "As Low As Reasonably Practicable," is a principle that arises from UK legislation (specifically the Health and Safety at Work etc. Act 1974). The Act requires "Provision and maintenance of plant and systems of work that are, so far as is reasonably practicable, safe and without risks to health." This principle is applied across all hazardous industries in the UK, allowing each organization and industry to determine their own most effective safety management system in the way that allows for economic viability (3). Specifically, the principle is defined in the three main categories shown in Figure 5.

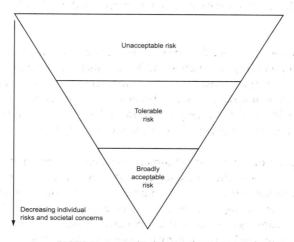

ALARP (As Low As Reasonably Practicable)

Figure 5. Illustration of the ALARP principle (4).

This information is intended to provide a high level reference to the ALARP principle. The reader is referred to the following website for a more definitive treatment of the topic as a basis for its use and application: (http://www.hse.gov.uk/risk/index.htm)

algae: Significant contributors to microbiologically influenced corrosion, algae can be either single-celled or multi-celled organisms that contain chlorophyll and generate oxygen by photosynthesis. The vast majority of algae are non-motile and reproduce by mitosis or sporulation. Algae can survive in a wide range of environments and their size varies enormously – all the way from micron size to large areas of kelp.

Algae can contribute to microbiologically influenced corrosion on pipelines both directly (by the production of organic acids from normal metabolic activity) and indirectly (by supporting other organisms, such as bacteria and fungi).

alignment sheet: A drawing, often overlaid with a photograph, that illustrates or shows the pipeline route, pipeline specifications, land ownership, topography, and installation information. An additional use of alignment sheets is as one of the inputs to the process for the location of in-line section anomalies.

alligatoring: Refers to pronounced wide cracking, having the appearance of alligator hide, over the surface of a coating that has oxidized and lost its ability to withstand mechanical/thermal expansion and contraction that is well within acceptable tolerances. Alligatoring is most often observed on tar- and asphalt-based coatings.

alloy: A substance made by melting two or more elements together, at least one of them a metal. An alloy crystallizes upon cooling into a solid solution, mixture, or intermetallic compound. Alloying is often undertaken to improve specific material properties without significantly adversely affecting others. One of the most commonly used alloys is brass. Steel is an alloy of iron and carbon. In the context of pipeline integrity, pipe steel has been alloyed to improve material properties and weldability.

alternating current (AC): A specific type of *electrical current* where the flow of electrons reverses direction at regular intervals. In North America, this cycling occurs 60 times per second (or 60 Hz) for most the of the power grid. This is in contrast to direct current where the flow of electrons is consistently in one direction (such as from batteries). From the point of view of the power grid, AC power experiences fewer transmission losses and necessary voltage changes are much cheaper to do.

From the point of view of non-destructive testing, AC current has very different characteristics (vs., DC) when traveling through metal (i.e., the current travels largely on the surface of the material vs., the entire cross section of the material) making it more suitable for certain applications such as detecting surface breaking cracks. AC also generates a correspondingly varying magnetic field – a phenomenon that has been explored and utilized in a number of non-destructive technologies.

alternating current voltage gradient (ACVG) survey: A type of survey that detects coating holidays and disbondments by measuring the rate of change of electrical potential of the ground in the vicinity of the pipe. The survey is very similar to a *direct current voltage gradient* (DCVG) except that it uses an alternating current power source instead of a direct current power source. See *Pearson survey*.

American Gas Association: Founded in 1918, the American Gas Association represents about 200 American energy companies that deliver natural gas throughout the United States Member companies deliver natural gas to more than 90% of the residential and commercial customers in the United States.

American National Standards Institute (ANSI): A non-profit organization that facilitates the development of American National Standards by accrediting the procedures of standards developing organizations. These groups work cooperatively to develop voluntary national consensus standards. Accreditation by ANSI signifies that the procedures used by the standards body in connection with the development of the American National Standards meet the Institute's essential requirements for openness, balance, consensus, and due process. ANSI is the US representative to the *International Standards Organization (ISO).*

American Petroleum Institute (API): The American Petroleum Institute is a US national trade association which represents all aspects of American oil and natural gas industry. It has about 400 corporate members, including the largest major oil company and smallest independents. The member companies come from all segments of the industry: producers, refiners, suppliers, pipeline operators, and marine transporters, as well as service and supply companies that support all segments of the industry.

Although the focus is primarily the oil and gas industry in the United States, recently it has expanded to include an international dimension. API activities include:
- Representation of the industry and member companies to governments, the public and regulatory bodies.
- Execution and sponsorship of research, as well as collection of statistics of interest to the oil and gas industry.
- Development of standards and recommended practices for petroleum and petrochemical equipment and operations.
- Support of certification and education.

American Society of Mechanical Engineers: Was established in 1880 by a small group of leading industrialists and is based in New York, New York. The organization is a wide-ranging technical community that offers continuing education, training and professional development, codes and standards, research, conferences and publications, government relations and other forms of outreach. Over the course of 130 years, ASME has grown to include more than 150,000 members in more than 120 countries across a range of industries that employ engineers. (192)

American Society for Metals: See *ASM International.*

American Society of Non-destructive Testing (ASNT): A non-profit technical society for non-destructive testing (NDT) professionals based in Columbus, Ohio. Through the organization and membership, ASNT provides a forum for exchange of NDT technical information; NDT educational materials and programs; and standards and services for the qualification and certification of NDT personnel. ASNT promotes the discipline of NDT as a profession and facilitates NDT research and technology applications.

ASNT was founded in 1941 (under the name of the American Industrial Radium and X-Ray Society). It currently has more than 10,000 members including more than 450 corporate partner affiliated companies. The Society is structured into local Sections (or chapters) throughout the world. There are more than 70 local sections in the US and 14 internationally. The membership represents a wide cross-section of NDT practitioners working in the oil and gas industry, manufacturing, construction, education, research, consulting, services, and the military.

anaerobic: A term used to mean "free of unbound oxygen" – usually used to describe an environment, process, or life form. An anaerobic environment is one where oxygen is not readily available for chemical processes. An anaerobic process is one that proceeds without the presence of oxygen. Anaerobic life (such as anaerobic bacteria) does not require oxygen to survive; various forms of anaerobic life can or cannot tolerate the presence of oxygen. From a pipeline integrity point of view, whether an environment is *aerobic* or anaerobic can be important in understanding (and mitigating) *microbiologically influenced corrosion.*

anaerobic bacteria: See *bacteria.*

ANSI: See *American National Standards Institute.*

angle-beam ultrasonics: An ultrasonic method to detect cracks in a material. The method uses ultrasonic probes mounted at an angle with the surface of the material. See Figure 6.

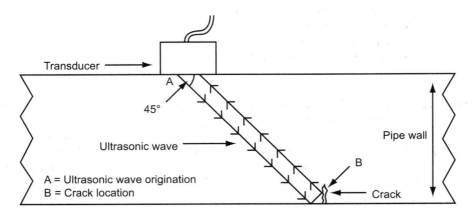

Figure 6. Angle-beam reflection from a crack. (1)

anion: A negatively charged ion (i.e., molecule or atom). In an electrochemical reaction, anions may exist in solution and are also produced at the cathode. These negatively charged ions migrate through the electrolyte from the cathode towards the anode. The migration of anions is illustrated in Figure 34. The movement of anions and cations is fundamental to the effectiveness of cathodic protection systems (see *Anode*).

annealing: A treatment consisting of heating a material to, and holding at, a suitable (i.e., high) temperature, followed by cooling at a suitable rate; used primarily to soften metallic materials or to relieve internal stresses, but also to simultaneously produce desired changes in other properties or in microstructure. Specific annealing temperatures and associated cooling rates are a function of the metal composition and the nature of the final microstructure desired.

annular mist flow: A form of *multiphase* flow where a liquid phase becomes entrained within a gas phase in the form of small droplets while a liquid film remains on the pipe wall. A schematic representing this flow regime appears in Figure 7.

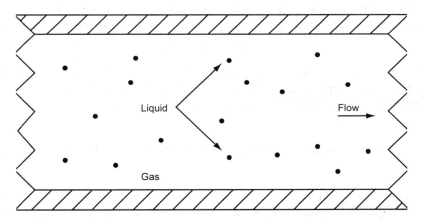

Figure 7. Schematic of mist flow. (10)

anode: See *electrochemical cell.*

anode cap: An electrical insulating material placed over the end of the sacrificial anode at the lead wire connection. The unit protects the connection from corrosion, which causes anode failure due to the detachment of the wire from the anode.

anode corrosion efficiency: The ratio of the actual corrosion (mass loss) of an anode to the theoretical corrosion (mass loss) calculated from the quantity of electricity that has passed between the anode and cathode using Faraday's Law. (5) The corrosion efficiency is obviously a consideration in cathodic protection system design as it indicates how long an anode can remain in service. The nature of any corrosion on an anode is not only a key factor in its corrosion efficiency, but can also play a critical role in affecting whether portions of the anode itself become detached, thus reducing the useful life of the unit.

anode suspension system: The use of a suspension system is a standard method of anode installation in the ground for *cathodic protection*. The anode is suspended by a cable and lowered into a hole in the ground. See *deep ground bed*.

anode utilization factor: A design consideration in cathodic protection design. Specifically, it is the percentage of an anode that can be consumed before the anode is no longer effective and requires replacement. Typically the anode utilization factor is about 85%.

anodic inhibitor: An inhibitor that reduces the corrosion rate by interfering with the corrosion reaction at the anode where the oxidation reaction occurs. The group of inhibitors based on imidazolines are generally anodic inhibitors. See *inhibitor*.

anodic polarization: The change of the electrode potential in the noble (positive) direction caused by current across the electrode/electrolyte interface. See *polarization*.

anodic reaction: The reaction at the anode where positively charged ions are transferred from the anode into the electrolyte. One of the most common reactions on pipeline steel is

$$Fe \rightarrow Fe^{2+} + 2e^-$$

In this reaction the anode is the source of the electrons, which are then carried to the cathode by the pipeline steel. The iron cations move into and through the electrolyte to complete the electric circuit. An anodic reaction is an oxidation process.

anolyte: A term for electrolyte in the immediate vicinity of the anode in an electrochemical cell.

anomaly: An unexamined possible deviation from sound pipe material or weld. An indication may be generated by any number of inferred measurement techniques such as an in-line inspection or x-ray. Often used interchangeably with the terms *indication* and *defect*.

ANSI: See *American National Standards Institute*.

AOPL: See *Association of Oil Pipelines*.

API: See *American Petroleum Institute*.

API gravity: The API gravity is a method developed by the American Petroleum Institute to describe the density of crude oil. The API gravity is calculated from the specific gravity:

$$API\ Gravity = \frac{141.5}{\rho} - 131.5$$

Where ρ is the specific gravity (density in gm/cm3)

A material with an API gravity of 10 has the same density of water. Most crude oil has an API between 10 and 70.

APIA: *See Australian Pipeline Industry Association*

arc strike: 1. A localized heat-affected zone, which may have a gouge-like appearance, caused by misplacing a welding electrode on the surface of the pipe. These features may prove to be an integrity concern because the localized heating and uncontrolled cooling of the pipe metal may have significantly degraded material properties, such as hardness, of the steel. As such, these features must be assessed on a case-by-case basis, and are typically removed by grinding.

2. API 577 3.5 arc strike: A discontinuity resulting from an arc, consisting of any localized remelted metal, heat affected metal, or change in the surface profile of any metal object.

arc welding: 1. A technique where heat in the weld area is introduced by causing/creating an electrical arc between an electrode and the base metal – it is one of the most common type of welding processes used in pipeline applications. The technique can be used with or without filler metal and may require shielding because the introduction of common substances (e.g., oxygen, hydrogen, nitrogen, and water vapor) reduces the quality of the weld) (6). The most common techniques used in pipeline applications are *Shielded Metal-Arc Welding* (*SMAW*), *Gas Metal Arc Welding* (*GMAW*), *Gas Tungsten Arc Welding* (*GTAW*), *Flux Cored Arc Welding* (*FCAW*), and *Submerged Arc Welding* (*SAW*). See also *welding*.

2. API 577 3.6 arc welding (AW): A group of welding processes that produces coalescence of work pieces by heating them with an arc, The processes are used with or without the application of pressure and with or without filler metal.

ASM International: Formerly the American Society for Metals, this is a society for the advancement of materials science and engineering. It began in 1913 as the Steel Treaters' Club. Since that time, it has grown to include about 36,000 members globally. The organization's mandate is to promote materials science and engineering through international conferences and expositions, seminars, local chapter meetings, and publications. Its publications include *Advanced Materials & Processes* and the *ASM Handbook* series.

ASME: See The *American Society of Mechanical Engineers*.

ASNT: See *American Society of Non-destructive Testing*.

asphalt enamel coating: Refers to a range of asphalt-based (i.e., derived from bitumen) external pipeline coating formulations. The asphalt is mixed with carbon to form a hardened physical barrier on the surface of the pipe when dry. See *coating*.

asphalt mastic coating: Refers to a range of asphalt-based (i.e., derived from bitumen) external pipeline coating formulations that contain various substances including sand, crushed limestone, and glass fibre. The coating tends to be heavy and costly to produce and has had limited applications. See *coating*.

asphalt tape: Refers to various asphalt-based (i.e., derived from bitumen) external pipeline coating formulations that are applied to the pipeline in tape form. Due to the shortcomings of tape coatings, this coating type has seen limited use in pipeline applications. See *coating*.

assessment criteria: See *assessment method*.

assessment interval: The time interval between successive reviews of a pipeline's condition. The assessments are typically based on *in-line inspection* data or by *direct assessment* techniques. Assessment intervals are typically set based on a combination of measured defect geometry as well as expected corrosion growth rate(s). While the calculations can be undertaken on a *deterministic* (i.e., non-probabilistic) basis, the use of probabilistic methods is increasingly common. Regardless of the method used to set assessment intervals, it should be noted that current thinking is that assessment intervals should likely not exceed 7 to 10 years given the sensitivity of results to growth rate assumptions – values that are extremely difficult to establish accurately.

assessment method: A means for determining whether damage to a pipeline due to corrosion, cracking or other flaws is injurious. In the context of pipeline integrity, assessment methods generally use the dimensions of the flaw, as well as pipe diameter, wall thickness, and grade to determine the safe operating pressure. That is, the burst pressure of any flaw that does not meet the relevant minimum threshold for the pipe characteristics, operating pressure and class of service, would require repair. In some pipes low toughness may be the governing factor.

Most assessment methods are approximations because they cannot account for complex geometry of the flaw or all properties of the material. Several assessment criteria have been developed with varying accuracy and conservativeness. Generally, the more accurate methods require more complex calculations. Table 4 lists some commonly used assessment criteria, used for corrosion and metal-loss damage.

Each of the assessment criteria calculates (either implicitly or explicitly) a predicted burst pressure. A repair is assessed if the following comparison is false:

$$MOP \leq P_{burst} \times FJLT$$

where MOP = maximum operating pressure

P_{burst} = predicted burst pressure

$FJLT$ = product of the design, joint, location, and temperature factors

Defect assessments may also be carried out for dynamic or cyclically applied loading conditions, in which case the load spectra and fatigue properties of the pipe material are required.

Table 4. Common assessment criteria for corrosion (8)

Assessment method	Description
B31G	Method developed in the early 1970s based on the NG18 equation. See *B31G*.
Modified B31G	Improvements over the B31G equations include Folias Factor, flow stress, and effective area of metal-loss. See *modified B31G*.
RStreng®	Based on the modified B31G equation but uses a procedure to find the critical sub defect within a larger complex area of corrosion. See *RStreng*.

Association of Oil Pipelines (AOPL): A non-profit organization started in 1947. As a trade association, AOPL:
- Acts as an information clearinghouse for the public, the media, and the pipeline industry.
- Provides coordination and leadership for the industry's ongoing Joint Environmental Safety Initiative.
- Represents common carrier crude and product petroleum pipelines in Congress, before regulatory agencies, and in the federal courts.

Australian Pipeline Industry Association (APIA): Originally started as a construction contractors association when it was established in 1968. The organization is an industry association managed through a central secretariat located in Canberra, Australia. Over the course of 40 years, APIA has grown to include more than 400 member organizations across the country and is very active both domestically and internationally as a representative of the Australian Pipeline Sector. (191)

automated NDE (or NDT) inspection: A non-destructive evaluation or testing method that is largely conducted by an automated inspection system. Typically, an automated inspection system is attached to the pipeline by belts, clamps, or magnets. The equipment automatically moves over the surface of the pipe, recording signals in a grid pattern over a set area of the pipe. An automated inspection is usually faster and more accurate than a manual inspection. The actual technique used to establish measurements will vary from system to system, but the most common methods include laser mapping and ultrasonic technology.

axial grooving: Refers to the approximate shape of long narrow metal loss anomalies, axially aligned, as identified through in-line inspection. Specifically, the dimensions of the anomaly are

$$width < length / 2$$

and

$$A \leq width \leq 3A$$

Where A = wall thickness if $t < 10$, else $A = 10$

This definition is based on a document published by the *Pipeline Operators Forum*; additional detail is provided in Figure 29 under *defect classification*.

axial sensor: A sensor mounted on a magnetic flux leakage inspection tool that measures the magnetic flux (or change in *magnetic flux* in the axial direction). The axial direction, used in a cylindrical coordinate system, is aligned with the z-axis or along the length of the pipeline (See Figure 8.). Magnetic flux tools typically have several sets of sensors around the circumference of the tool. Each set, typically contains axial and radial sensors; some tools also have a circumferential sensor.

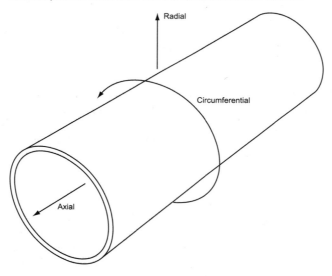

Figure 8. Axial, radial, and circumferential directions on a pipeline.

axial slotting: Refers to the approximate shape of long narrow metal loss anomalies, axially aligned, as identified through in-line inspection. Specifically, the dimensions of the anomaly are

$$width < A$$

and

$$length > A$$

where A = wall thickness if $t < 10$, else $A = 10$

The definition is based on a document published by the *Pipeline Operators Forum*; additional detail is provided in the Figure 29 under *defect classification*.

axial stress: In a *cylindrical coordinate system*, axial stress is located in the longitudinal direction, that is along the length of the pipeline. While this equation applies to a closed system where the fluid is static (e.g., hydrotest), it is considered approximately correct for the case where fluid is moving through the line.

B

B-scan: One of the presentation formats used to display ultrasonic testing results where signal amplitude is plotted against either the length of width of the component (as per Figure 9. B-scan display.). The time of the reflection (y-axis) is directly related to the thickness of the sample under investigation. The x-axis is the location of the sensor as it is moved in one direction along the surface of the specimen. Each point on the B-scan represents a separate reading. See also *A-scan* (plot of signal amplitude vs. time) and *C-scan* (representation of apparent corrosion on a two-dimensional view of the component.

The B-scan is constructed from a series of A-scan displays. The time-of-flight is the time until the first signal (after the initial pulse) exceeds a pre-determined threshold amplitude. The time-of-flight is then plotted on the vertical axis of the B-scan.

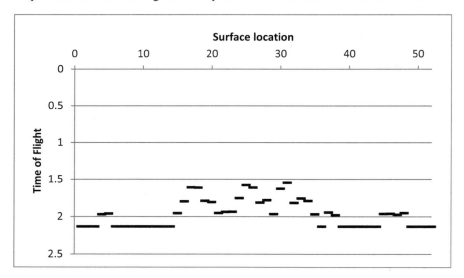

Figure 9. B-scan display.

B31G equation: This equation is found in the B31G supplement to the ASME B31 Code and is used to calculate the burst pressure of a non-indented blunt axial flaw, such as corrosion, in a pipeline.

The equation is (8):

$$P_f = \frac{2St}{D} \frac{(1 - \tfrac{2}{3}d)}{(1 - \tfrac{2d}{3M})}$$

where P_f = predicted failure pressure
S = the flow stress
S = 1.1 $SMYS$
$SMYS$ = specified minimum yield strength of the steel
t = wall thickness of the pipe
D = diameter of the pipe
d = depth of the corrosion
M = Folias Factor, which is further calculated by

$$M = \begin{cases} \sqrt{1+0.8\dfrac{L^2}{Dt}} & if \ \dfrac{L^2}{Dt} \leq 20 \\ \\ 1 & if \ \dfrac{L^2}{Dt} > 20 \end{cases}$$

where L = length of the corrosion

The B31G equation has been used widely but has been proven to be conservative (i.e., it predicts a burst pressure that is on average less than the actual burst pressure based on burst testing). Further, the variance of the ratio of predicted burst pressure to actual burst pressure is relatively large. As a result, a modified version of the equation ("modified B31G") was developed to address these difficulties:

$$P_f = \frac{2St}{D} \frac{(1-0.85d)}{(1-0.85\,d/M)}$$

where P_f = predicted failure pressure
S = flow stress
S = 10,000 psi + $SMYS$
$SMYS$ = specified minimum yield strength of the steel
t = wall thickness of the pipe
D = diameter of the pipe
d = depth of the corrosion
M = Folias Factor

$$M = \begin{cases} \sqrt{1+0.6275\dfrac{L^2}{Dt}-0.003375\dfrac{L^4}{D^2t^2}} & if \ \dfrac{L^2}{Dt} \leq 50 \\ \\ 0.032\dfrac{L^2}{Dt}+3.3 & if \ \dfrac{L^2}{Dt} > 50 \end{cases}$$

where L = length of the corrosion

The difference between the two equations is in the assumptions used to derive the relationships. A comparison of the key assumptions in the two equations appears in the following table:

Table 5. Comparison of B31G and modified B31G equation assumptions (8)

Assumption	B31G	Modified B31G
Maximum defect depth	80% of nominal wall thickness	80% of nominal wall thickness
Flow stress	1.1 x SMYS	SMYS + 10 "ksi" (68.95 Nmm⁻²)
Folias factor	Two-term Folias Factor	Three-term Folias Factor
Defect shape	-Parabolic defect shape for short (M ≤ 4.12); -Rectangular shape for long corrosion (M ≥4.12)	-Arbitrary shape assumption of 0.85 x length x max depth

Figure 10 shows the difference in assumptions between the two equations.

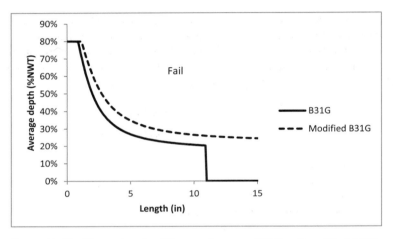

Figure 10. Plot illustrating the difference between B31G and modified B31G equations.

back pressure: The pressure exerted on the downstream side of a valve seat, or on the downstream side of an inspection tool. The difference between the upstream and downstream pressure (i.e., differential pressure) indicates the size of the forces that the valve must withstand in a given situation. In the case of an in-line inspection tool, the differential pressure will influence the tool speed while it is in the pipeline.

backfill: Any material placed in the ditch to fill the trench up to the normal grade. The material used as backfill can range from previously excavated soils to specially ordered materials (such as sand) to ensure an extra degree of protection for the pipe. The term is also used in the context of the installation various elements of cathodic protection systems (such as anodes, vent pipe, etc.,) where material is needed to fill the drilled hole.

The nature of the backfill can affect pipeline integrity as it can significantly contribute to conditions leading to cracking and corrosion. For example, some older pipeline construction practices deemed it acceptable to use previously excavated rock as backfill – ultimately leading to coating damage and a greater risk of corrosion failures.

bacteria: One of the most significant contributors to microbiologically influenced corrosion. Bacteria are simple single-celled organisms that are usually less than 1 μm in diameter but can be as large as 10 μm. Bacteria are very versatile and live in a wide range of temperatures, pH levels, and oxygen levels. In some cases, in unfavorable conditions, they become dormant and form endospores. They may remain dormant for extended periods of time. When conditions improve, they become active again.

Bacteria pose a threat to pipelines by their influence on both internal and external microbiologically influenced corrosion (MIC). Bacteria associated with MIC include:
- Sulfate-reducing bacteria
- Sulfur-sulfide-oxidizing bacteria
- Iron/manganese-oxidizing bacteria
- Aerobic slime formers
- Methane producers
- Organic acid-producing bacteria

ball valve: A spherical closing element held between two seats. They are widely used in applications that require quick opening/closing with good shut-off properties. There are two main types of ball valves. Figure 11 shows a schematic of a ball valve, and Table 6 describes the differences between the two main types.

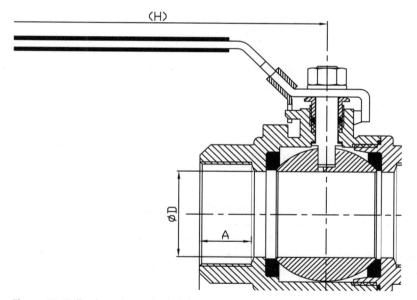

Figure 11. Ball valve schematic. (200)

Table 6. Major subtypes of ball valves (27)

Subtype	Characteristics
Floating ball	Small diameter balls that do not require internal support.
Trunion mounted ball	Heavier balls (typical in pipeline applications) require the use of trunions to support the ball and prevent damage to soft internals.

Barlow's formula: A formula used to calculate the hoop stress due to internal pressure in a thin-walled cylinder or pipeline. The hoop stress, σ_θ, is given by the following formula:

$$\sigma_\theta = \frac{PD}{2t}$$

where P = internal pressure
D = diameter of the pipeline
t = wall thickness

Barlow's formula is also used to calculate the maximum allowed pressure in a pipeline given a safe hoop stress.

$$P = \frac{2St}{D}$$

where S = maximum allowed hoop stress in the material (generally between the specified minimum yield strength and the ultimate tensile strength of the material).

Barnes method: A procedure to calculate a profile of soil resistivity as a function of depth – a key input parameter in the design of cathodic protection systems.

barred tee: A tee fitting that has cross bars on the transverse opening. Barred tees can pose a hazard to pigging operations if they are not flush with the internal surface of the pipeline – further a tee must be barred for pigging operations if it branches off a sufficiently large diameter (i.e., > 60% of the main pipeline diameter) – especially if it connects to the main pipeline at less than a 90° angle. See *tee fitting*.

batching: Sequential transportation of liquids to transport multiple products through a single pipeline. This is sometimes accomplished by "enclosing" the fluid between a series of *batching pigs* to ensure separation from other products. Alternatively, and more often, a short length of synthetic fluid is introduced between batches.

Batching operations are used primarily for two purposes: transporting two or more products in the same pipeline, and applying *corrosion inhibitors* and *biocides*. In some cases ultrasonic tools have been run in gas lines contained in a batch of liquid to provide an effective medium for transmitting the ultrasonic signal. It should also be noted that the starting and stopping associated with batching can cause cyclic stresses leading to fatigue-based failure mechanisms.

batching pig: A utility pig used in *batching* to separate two different fluids or batches flowing in the same pipeline. See *pig*.

bathtub failure curve: This curve is a graphic description of the nature of how failure rates for many types of engineering devices vary over time. The curve consists of three parts, as illustrated in Figure 12:

1. An early period, with a high rate of failure due to manufacturing, construction, and material flaws (e.g., infant mortality).
2. A middle period, with a low rate of failure due to random events. (e.g., normal life).
3. A late period, with a rising rate of failure due to the wear-out (e.g., end of life).

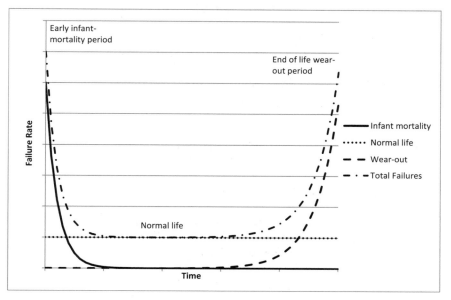

Figure 12. Bathtub failure curve.

beach marks: Characteristic markings on the fracture surfaces produced by fatigue crack and SCC propagation (also known as clamshell marks, conchoidal marks, and arrest marks).(5) Macroscopic progression marks on a fatigue fracture or stress corrosion cracking surface that indicate successive positions of the advancing crack front. The classic appearance is often irregular elliptical or semielliptical rings, radiating outward from one or more origins.

bead pass: A layer of weld material that is laid down without appreciable side-to-side oscillation of the welding electrode. See *weld passes*.

bell-hole: A short trench exposing the pipeline to permit survey, inspection, maintenance, repair, or replacement. While there is no industry standard definition regarding trench dimensions, the term is typically used to describe excavations just long enough to expose both girth welds; i.e., a trench a little greater than a pipe joint (36 ft/12 m).

bell-hole examination: The direct examination of the pipeline in a ditch of limited length (i.e., typically slightly more than one joint of pipe) – that is, inspection of the length of pipe exposed in a *bell-hole*.

berm: Any earth or rock structure (i.e., mound) used to divert surface water flow and provide slope stability in geotechnically sensitive areas (8). Berms are also used in various other applications, such as containment around (liquid) storage tanks in the event of tank failure.

beta distribution: A probability distribution defined on the interval [0,1]. Mathematically, the probability density function (PDF) of the beta distribution, $beta(x)$, is defined as:

$$beta(x) = \begin{cases} \dfrac{\Gamma(\alpha+\beta)}{\Gamma(\alpha)\tilde{A}(\beta)} x^{\alpha-1}(1-x)^{\beta-1}, & 0 \leq x \leq 1 \\ 0, & x < 0 \, and \, x > 1 \end{cases}$$

where α and β = shape parameters of the distribution

$\Gamma(\alpha)$ = Gamma function

$$\Gamma(\alpha) = \int_0^\infty \left(\frac{x}{\beta}\right)^{\alpha-1} e^{-x/\beta} \frac{1}{\beta} dx$$

Table 7 shows selected properties of the beta distribution.

Table 7. Beta distribution properties

Parameters	α and β
Domain	$0 \leq x \leq 1$
Mean	$\dfrac{\alpha}{\alpha+\beta}$
Variance	$\dfrac{\alpha\beta}{(\alpha+\beta)^2(\alpha+\beta+1)} \alpha\beta^2$
Mode	$\dfrac{\alpha-1}{\alpha+\beta-2}$

Beta distributions are often used to model estimates of ratios or probabilities. Specifically, they are often used in the *quantitative risk* based analysis of in-line inspection (in-line inspection) data. An estimate of the proportion of in-line inspection depths that are within tolerance can be modeled as a beta distribution. Beta distributions have also been used as the probability distribution of the true depth given a measured corrosion depth. Figure 13 shows selected samples of beta distributions, demonstrating the general shape of the function.

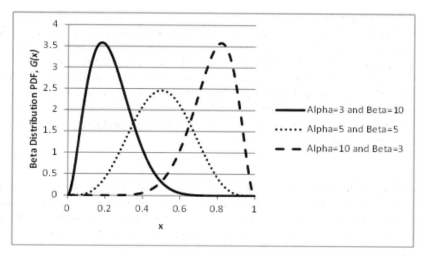

Figure 13. Beta distribution for range of α and β. The exponential distribution is where α=1.

bi-di pig: See *bi-directional pig.*

bi-directional pig: A utility pig mounted with discs (instead of cups) to facilitate movement in both directions within a pipeline. See pig.

bimetallic corrosion: See *galvanic corrosion.*

binder: Generic term for the substance in a pipeline coating system that allows particles to adhere to one another. One of the most common examples is the resin base used in fusion bond epoxy coatings. See coating.

biocide: A chemical that can kill or inhibit the growth of living organisms such as bacteria, fungi, molds, and slimes. Because they inhibit living organisms, most of them are also toxic, to some degree, to humans. As such, biocides typically require specific handling procedures. They are often used in pipelines to control several organisms (microbiologically influenced corrosion) that cause internal corrosion. Application of biocide can be done either continuously on-stream or done as a batch. Table 8 shows the most common types of biocides.

Table 8. Major types of biocides used in pipeline applications (18)

Chemical Group	Mechanisms
Aldehydes	• A class of fixatives that crosslink the cell wall material so that nutrients cannot enter the cell; the cell "suffocates."
Quaternary compounds	• A class of surfactants that break down the cell wall structure so that cellular fluids leak out; the cell "bleeds to death."
Phophonium compounds	• A class of pesticides that interfere with cellular metabolism; the cell is essentially poisoned.

biofilm: A layer of organic material covering a surface. Biofilms are important in the context of pipeline integrity because they relate specifically to microbiologically influenced corrosion (MIC). Biofilms are composed of various microorganisms including algae, fungi, and bacteria. The films also contain biopolymers that are produced by aerobic slime-forming bacteria. The biopolymers help the microorganisms to stick to the surface. The flow rate in the pipeline does not prevent the formation of a biofilm, but it will affect the structure of the film. Biofilms affect corrosion in several ways including differential aeration, increasing acidity at the pipe surface, and the creation of an environment for sulfide-reducing bacteria.

biofouling: The accumulation of biological organisms on a component surface. In general, biofouling can refer to the accumulation of microorganisms, such as bacteria, as well as larger organisms, such as seaweed and barnacles. On pipelines, biofouling refers to the formation of a *biofilm* that has significant implications for *microbiologically influenced corrosion*.

bituminous coating: Various asphalt-based external pipeline coating formulations that include *asphalt enamel, asphalt mastic,* and *asphalt tape.* See *coating*.

black powder: A range of iron-type products that collect in gas pipelines. The substance is largely a mixture of iron sulfide and iron oxide. Sufficient quantities of black powder may interfere with the operation of valves, meters, and pipe flow. A goal of some pipeline cleaning operations is to remove the accumulation of black powder. Further, there is a significantly greater risk of fire when black powder is exposed to oxygen and as such, special precautions may be necessary when receiving pigs (both cleaning and smart tools) that have been exposed to significant amounts of black powder.

blister: **1.** A coating blister is a raised area on the coating, often dome-shaped, resulting from either loss of adhesion between a coating and the base metal. **2.** A metal blister is a raised area, often dome-shaped, resulting from localized delamination or void in the metal. The delamination is the result of an expanding gas, such as hydrogen, trapped in a metal in a near-subsurface zone. See also *hydrogen damage.*

block valve: See *main-line valve.*

blowdown: **1.** The controlled evacuation of pressurized gaseous product from equipment or facilities for various pipeline operations including but not limited to emergency station shutdown, launching and receiving pigs, and preparation for pipeline repair. **2.** Injection of air or water under high pressure through a tube to the anode area for the purpose of purging the annular space and possibly correcting high resistance caused by gas blockage. **3.** In conjunction with pressure vessels or pipelines, the process of discharging a significant portion of the liquid product to remove accumulated salts, deposits, and other impurities.

blowdown valve: A valve used in *blowdown* operations. Where blowdown is part of emergency shutdown procedures, the valves are automated.

blusing: Refers to the visual manifestation, in the form of whitening and dulling, of an organic coating that has absorbed moisture. It is used interchangeably with the term blooming.

booster station: The set of pumps or compressors and associated facilities enclosed within a station yard, used to propel liquid product down the pipeline.

box: This term is used in the context of magnetic flux leakage or ultrasonic data from in-line inspection tools. Specifically, a box is a metal-loss anomaly that has been identified by the in-line inspection tool. The size of a box is given by a length, width, and depth. Unless the box is sufficiently isolated from other metal-loss anomalies, adjacent box anomalies must be considered to calculate burst pressure see *interaction rule* and *cluster*.

bridging bar: A device used to provide an appropriate reference point when taking corrosion depth measurements in an area of extensive metal loss on a pipeline. A bridging bar consists of a bar with supports at both ends. The technician rests the bridging bar on its supports, axially aligned, and ideally on uncorroded pipe, over the corrosion anomaly. The depth of corrosion is then measured using a pit gauge mounted on the bar suspended between the supports.

Brinnell hardness: A standardized non-destructive test used to determine the hardness of a material. Hardness is a measure of a materials resistance to plastic deformation or penetration and is an important parameter to control and assess on a pipeline because harder materials are often more susceptible to various forms of cracking. The Brinnell harness test estimates hardness by subjecting the surface of a material with an indenter of a specific size and shape and with a specific force. The hardness is estimated by examining the size of the indentation and can be related to *ultimate tensile strength* through an empirically determined equation. See *hardness* for a comparison of major test methods.

brittle fracture: In a structural engineering context, describes the failure of a component when there is little or no plastic deformation. Unlike *ductile facture,* brittle fracture occurs when the material breaks apart, like glass, without deforming the material. Typically, brittle fracture occurs by rapid crack propagation with much less expenditure of energy than for *ductile fracture*. See also *fracture control*.

brush pig: A utility pig adapted for a specific pipe wall cleaning application through the addition of brushes. See *pig*.

brush-off blast cleaned surface: A term to describe the level of cleaning of a surface. A brush-off blast cleaned surface, when viewed without magnification, is free of all visible oil, grease, dirt, dust, loose mill scale, loose rust, and loose coating. Tightly adherent mill scale, rust, and coating may remain on the surface. Mill scale, rust, and coating are considered tightly adherent if they cannot be removed by lifting with a dull putty knife.

brush-wheel pig: A specific type of utility pig used for cleaning of pipelines. Small circular brush pods are mounted on a body (typically constructed from a material such as urethane). The brushes radiate out from the pig to contact the internal surface of the pipe. Brush-wheel pigs effectively clean the sides and bottoms of large corrosion pits. Depending on the specific nature of the cleaning operations, the brush-wheel pig may need to be coupled with additional tools to ensure sufficient debris removal occurs and/or differential pressure to keep the tool moving. Brush-wheel cleaning pigs are often run as part of a cleaning program to prepare for an in-line inspection run or prior to the application of an inhibitor batch. See also *pig*.

bubble flow: A form of *multiphase* flow where a gas phase becomes entrained within a liquid phase in the form of bubbles. Figure 14 shows a schematic representing this flow regime.

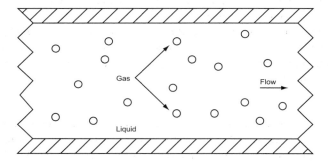

Figure 14. Schematic of bubble flow. (10)

buckle: The result of compressive mechanical damage, usually caused by ground movement, on a pipeline. Specifically, it is a situation where the pipeline has undergone sufficient plastic deformation to cause a permanent wrinkling or deformation of the pipeline cross-section. Obviously buckles represent a problem since structural integrity of the pipeline has been compromised; further, there are additional risks resulting from restricted pig passage, coating damage and increased susceptibility to cracking of the buckled surfaces.

buffing: See *grinding*.

bulk modulus: An elastic constant of materials and fluids, which defines the relationship between an applied pressure and the resulting change in volume – as such, it is one of the parameters used in several equations to estimate failure pressure. The bulk modulus, K, is the ratio of the applied pressure to the change in volume of a unit volume.

$$K = -\frac{P}{e}$$

where P = applied pressure
e = resulting change in volume per unit volume

The bulk modulus is also related to other moduli such as:

$$K = \frac{E}{2(1-2v)}$$

where E = Young's modulus
v = Poisson's ratio

Figure 15 illustrates the stress and strains relationships for the bulk modulus.

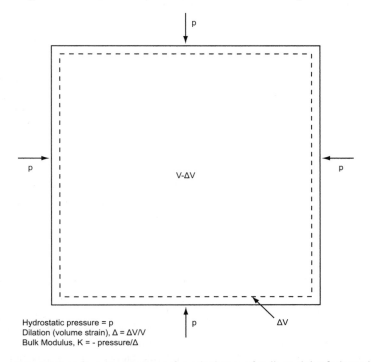

Hydrostatic pressure = p
Dilation (volume strain), Δ = ΔV/V
Bulk Modulus, K = - pressure/Δ

Figure 15. Relevant parameters for calculation of Bulk modulus [adapted from (10)].

burn marks: Localized areas on either side of the seam weld where the pipeline steel has been altered by the welding process. Burn marks can be associated with lack of fusion in the seam weld.

burn radius: Refers to the land near a pipeline expected to experience flammable conditions in the case of rupture and ignition of an oil or gas pipeline. The radius is a function of the specific composition of the pipeline product, the diameter, as well as operating pressure and temperature. This term is more commonly referred to as the *potential impact radius*.

burnthrough: 1. A weld flaw where the first weld pass melts through, thereby creating a root flaw. Often these are difficult to repair. **2.** In a mechanized weld utilizing an internal root pass, a burnthrough may occur during the first pass from the OD (hot pass) if the root bead is not present. **3.** During in-service welding, a burnthrough may occur if the remaining wall is too thin for the level of heat input. **4.** A welding feature where the root bead is absent, and can affect the integrity of the pipeline. The root bead is lost when the unmelted metal has insufficient strength to maintain the internal pipeline pressure, during the welding on a hotline. The risk and occurrence of burnthrough is a function of the pipe wall thickness, operating conditions, and welding parameters.

burst: A pipeline failure mode in which there is a large release of product because the pipe is unable to contain the internal pressure and the pipeline fails by a combination of ductile and brittle fractures – however, there is no *running fracture*. It should be noted that usage is not consistent and at times the term is used interchangeably with *rupture*; the other loss-of-containment failure mode is *leak*.

burst pressure: The pressure at which a pipeline is expected to fail in a sudden, uncontrolled manner (i.e., *burst*). Burst pressure can be calculated by several different methods with varying degrees of accuracy and conservatism. Common burst-pressure calculation methods include *B31G, modified B31G, RStreng*, and elastoplastic *finite-element modeling*.

burst test: A test where a short length of pipeline is purposefully made to burst. The procedure includes welding end caps onto the pipe and pumping air or water into the pipe section. The pressure is usually closely monitored and recorded to determine at exactly what pressure the pipe fails. The purpose of a test is to test pipe materials, the effectiveness of a repair, or the accuracy of a burst pressure calculation.

butterfly valve: This valve uses a rotating disk, requiring a quarter turn, which interrupts the flow of fluid in a pipeline. Depending on the specific design of the valve, it can be used for shut-off and throttling applications. Most standard designs have the valve stem running through the disk, giving a symmetrical appearance. However, designs with an offset valve stem are also available and tend to have better shut-off capabilities. Butterfly valves do pose a hazard to pigging operations because the disk remains in the flow of the product when it is open. Figure 16 shows the basic form of a butterfly valve.

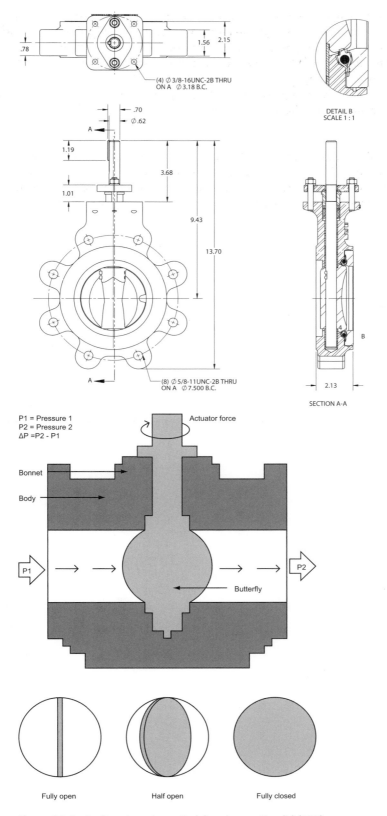

Figure 16. Butterfly valve schematic (a) and operation (b) (200).

bypass pig: A *pig* designed to allow some pipeline product to bypass the sealing cups. This feature is not a standard configuration for most pigs and is not specific to any particular type of pig (i.e., cleaning pig, smart pig). The use of bypass is primarily for situations where product flow is faster than the optimal tool speed. Further, the nature of the bypass system can vary greatly – anywhere from small holes in the driving cups of a cleaning tool to a very sophisticated valving system that monitors and controls tool speed to facilitate the collection of magnetic flux leakage data.

bypass valve: A valve that is fitted in parallel to a larger main valve. Bypass valves are used to reduce the differential pressure across the main valve before the latter is opened to prevent this larger, more expensive valve from being damaged. Depending on the diameter of the bypass, the bypass line maybe used to temporarily redirect flow.

C

C-scan: A presentation format used to display ultrasonic testing results where the size and location of features are identified through color on a plan view of the specimen. The x and y axes of the C-scan match the x and y axes on the surface of the specimen. At each x-y location, the time of flight, signal amplitude, or thickness is posted or color coded at the corresponding position on the C-scan display. See also *A-Scan* (plot of signal amplitude vs. time) and *B-scan* (plot of signal amplitude vs. one-dimensional view (such as length or width) of the component.

calcareous: Coatings that are found on cathodically protected surfaces as a result of increased pH adjacent to the protected surface.

calibration dig: A calibration dig is an excavation aimed (in part) at locating and examining an anomaly that was reported in an in-line inspection report, for the purpose of comparing the in-line inspection reported characteristics to those found in the field. Specifically, the pipeline is exposed and specific anomalies are located, sized and compared to the results from an internal inspection tool. Historically, the term arises out of a need in the past to calibrate the sizing algorithms with the excavation results. Today, the term is synonymously with *verification dig*. To serve as a verification dig, generally, a sufficient number of anomalies must be assessed to ensure that the results are statistically valid prior to reaching a conclusion regarding tool accuracy. The data from the dig is then used as part of the process of correcting for in-line inspection tool measurement error. See also *verification dig*.

caliper tool: A type of smart pig used to assess the nature of the pipeline geometry and sizing – usually prior to undertaking an inspection for corrosion or cracking. The term is used interchangeably with *geometry tool* and *deformation tool*. See *pig* for a detailed discussion of this class of tools.

camera pig: A free-swimming (i.e., non-tethered) camera tool. See *camera tool*.

camera tool: A tool capable of photography or videography, either intermittently or continuously, as a means of inspecting the inside of the pipeline. The tools may take various forms – the specifics of which will vary depending on the situation. For example, tethered tools may be best suited for smaller diameters or short pipeline sections whereas longer sections may require a free swimming or "pigging" tool.

Canadian Energy Pipeline Association (CEPA): This association represents Canada's transmission pipeline companies. According to CEPA, members provide safe, reliable long-distance energy transportation. Transmission pipelines transport nearly all of Canada's daily crude oil and natural gas production from producing regions to markets throughout Canada and the United States.

Canadian Standards Association (CSA): A not-for-profit membership-based association serving business, industry, government and consumers in Canada and the global marketplace. The CSA provides engineering standards to the Canadian pipeline industry.

cap: The final layer of weld material deposited during a welding operation. See *weld passes.*

cap undercut: A type of girth weld anomaly where the parent pipe material has melted away, thereby creating a smooth groove in the parent metal adjacent to the top bead of the weld cap. See Figure 17.

A gouging out of the piece to be welded, alongside the edge of the top or "external" surface of the weld.

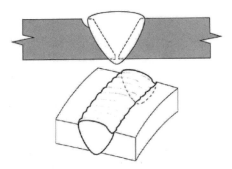

Figure 17. Cap undercut. (197)

carbon backfill: A commonly used material to surround an anode within its ground bed to improve hole integrity and ensure good electrical conductivity.

carbon dioxide: A gaseous compound that is present in the atmosphere, soil, and in many natural gas reservoirs. Carbon dioxide also is produced by the combustion of carbon compounds, plants, and by organic decay. The chemical formula is CO_2.

In pipelines, carbon dioxide can cause internal corrosion when it is contained in the gas or liquid pipelines. Carbon dioxide may be a natural component of the product, but it may also be introduced by secondary and tertiary recovery methods. Carbon dioxide can often participate in and accelerate corrosion reactions – especially in the presence of water because it forms carbonic acid (see *CO_2 corrosion*). Carbon dioxide is also a common component of organic-rich soils and is thought to increase the likelihood of the initiation and growth of *stress corrosion cracking* (SCC).

Increasingly, carbon dioxide is transported in pipelines as a product. As a product, it poses several unique challenges to pipeline integrity, including but not limited to increased risk of internal corrosion in the presence of water, shortened life of seals in equipment such as valves, significant toxicity in large volumes of product (in case of pipeline failure), as well as increased risk of running fracture, because carbon dioxide is a liquid at typical pipeline temperatures and pressures. Another significant difference in transport of pure vs. impure CO_2 is the latter needs much higher pressure, and if the line leaks, the Joule-Thomson effect can lead to low-temperature cracking.

carbon steel: A generic term for an alloy of iron and carbon. Carbon steels are defined as steels with no specified minimum content for aluminum, boron, chromium, cobalt, molybdenum, nickel, niobium, titanium, tungsten, vanadium, zirconium or any other element. Carbon steels do have a maximum content for some elements, as shown in Table 9.

Table 9. Composition specifications for carbon steels (195)

Element	Maximum content
Copper	0.60 %
Manganese	1.65 %
Silicon	0.60%
Carbon	0.05 – 2.0%

Further, carbon is added to the steel to harden and strengthen it. As the carbon content increases the hardness and strength of the steel increases, while toughness is reduced. Low carbon steel, with a carbon content of 0.001 % - 0.3 % by weight, is a commonly used material in the construction of pipelines.

casing: A protective metal cylinder surrounding the pipeline, usually made of larger diameter pipe, installed for the purposes of protecting the pipeline from external damage. Casings are often used under railroads, highways, roads, and streets, where stresses from the overlying road may otherwise damage the pipeline.

Casings can create problems if they are shorted (electrically connected) to the pipeline. The shorted casings will shield the cathodic protection from the pipeline, leaving it unprotected from external corrosion if the coating fails. Casings can also act as a vehicle for the pooling of water along the pipeline.

Category I-IV SCC: See *stress corrosion cracking.*

cathode: See *electrochemical cell.*

cathodic disbondment: The disbondment of a coating from the surface of a metal due to a cathodic reaction at the surface of the metal. Specifically, the cathodic protection levels may be higher than generally accepted, resulting in coating damage. A common mechanism blamed for cathodic disbondment is the evolution of hydrogen. The cathodic half-cell reaction is:

$$2H^+ + 2e^- \rightarrow H_2$$

The production of molecular hydrogen on the external surface of a pipeline is thought to cause the coating to blister and disbond. However, this theory is not universally supported.

cathodic inhibitor: A chemical substance that prevents or reduces the corrosion rate by reducing the rate of the cathodic or reduction reaction. See *inhibitor*.

cathodic polarization: The change in potential voltage, in the active (negative) direction that results from the presence of current flowing. See *polarization*.

cathodic protection (CP): This is a widely-used method of electrically protecting a metal structure from corrosion. The usefulness of cathodic protection to protect metal from corrosion has been known since 1824. Its extensive use on steel pipelines began in the 1920s; however, until the 1940s, pipelines were generally constructed without cathodic protection.

Cathodic protection systems are effective because they make any natural (i.e., unprotected) sites that are active, more passive. More specifically, the reaction at a relatively active site would normally be as follows:

$$2 \text{ Fe} \rightarrow 2 \text{ Fe}^{2+} + 4 \, e^-$$

The free electrons then travel to the more passive site to participate in the cathodic reaction. For example, if oxygen is available, the electrons will combine with the oxygen and water:

$$O_2 + 4e^- + 2 \text{ H}_2O \rightarrow 4 \text{ OH}^-$$

The resulting products then recombine at the active site producing corrosion product:

$$2 \text{ Fe}^{2+} + 4 \text{ OH}^- \rightarrow 2 \text{ Fe (OH)}_2$$

A cathodic protection system then interrupts these chemical processes by providing a more readily accessible source of free electrons. As long as the rate at which electrons from the cathodic protection system are available is greater than the rate at which oxygen is available, the system will effectively protect the pipe.

There are two main types of cathodic protection: *galvanic* and *impressed current* systems. Galvanic cathodic protection, also known as sacrificial cathodic protection, relies on a metal such as zinc or magnesium to act as the (sacrificial) anode. Impressed current cathodic protection systems polarize the structure to be protected by an external direct current (DC) electrical power supply. See Figure 18 for a visual representation of the difference between the two systems.

Due to their fundamentally different natures, there are both advantages and disadvantages to both types of cathodic protection systems. The key differences are itemized in Table 10. The differences in galvanic vs. impressed current systems must then be assessed in light of four main criteria used to assess the effectiveness of a cathodic protection system as summarized in Table 11.

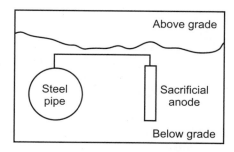
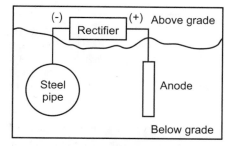

A) Galvanic system   B) Impressed current system

Figure 18. Galvanic vs. impressed current cathodic protection systems. (195)

Table 10. Comparison of galvanic and impressed current
cathodic protection systems (186)

Characteristic		Galvanic system	Impressed current system
Design / performance	Voltage range	Fixed (not adjustable)	Voltage adjustable over wide range
	Voltage size	Small voltage	Designed to meet needs
	Current range	Fixed (not adjustable)	Current adjustable over wide range
	Current size	Small current	Designed to meet needs
	Stray Current	Usually not a concern	Significant concern
Costs	Installation cost ($/unit)	Small	Large
	Maintenance cost	Low	High
	Unit cost ($/ft² protected)	Large	Small

Table 11. Four main cathodic protection criteria (5)

Criterion	Description	Application	Limitations
−850 mV applied	• Measurement taken while CP system is on relative to a reference electrode; specifically, in the case of a buried steel or cast iron structure, the criterion is met if the potential difference between the structure and a copper-sulfate electrode contacting the soil directly above and as close as possible to the structure is equal to or more negative than (larger in absolute value)−850 mV • Voltage drops other than those across the structure-to-electrolyte boundary must be considered; significance of measurements must be considered in context of: • measuring or calculating the voltage drop(s) • reviewing the historical performance of the CP system • evaluating the physical and electrical characteristics of the pipe and its environment • determining whether or not there is physical evidence of corrosion	• Most commonly applied criterion to coated and buried or submerged steel or cast iron structures; generally, sources of DC current not always easily interrupted in urban areas or if sacrificial CP system is in place	• Placement of electrode critical (directly over structure to minimize ohmic voltage drop errors and minimize extent of averaging over large areas of the structure) otherwise alternative criteria may be required • Most commonly used for well coated structures where it can be economically be met • Close interval surveys may be needed because potentials can vary significantly by location and by season (limitation of all criteria) • Specific conditions (e.g., hot pipelines or presence of microbes) may require a higher potential (i.e., −950 mV); however, overprotection should be avoided (general consensus in the industry is to avoid polarized (instant off) potentials more negative than−1.05 to−1.1 V) • Presence of stray currents complicates obtaining and interpreting survey data
-850 mV polarized potential	• Measurement taken immediately after current interruption (instant off) relative to a reference electrode	• Most commonly applied to coated structures where the sources of DC current can be readily interrupted	• In addition to limitations listed for first criterion, all DC currents must be interrupted

Table 11 *continued*

Criterion	Description	Application	Limitations
-850 mV polarized potential	• Method eliminates sources of IR drop inherent in first criterion – shift entirely a result of application of CP		
100 mV polarization	• Measurement of formation or decay of polarization relative to reference electrode in contact with electrolyte (i.e., over time once system has stabilized). • Of the three main criteria, this criterion has the most sound basis • The application of the 100mV polarization criterion has the advantage of minimizing coating degradation and hydrogen embrittlement, both of which can occur as a result of overprotection	• Commonly used on poorly coated or bare structures where it is difficult or costly to achieve either of the–850mVcriteria. • Because of its fundamental underpinnings, the 100 mV polarization criterion also can be used on metals other than steel, for which no specific potential required for protection has yet been established.	• Time required for full depolarization of poorly coated / bare structure can be several weeks • Complicated measurements mean high survey costs • Should not be used in areas of stray current • All DC current sources must be interrupted • Not applicable for structures that contain dissimilar metal couples or where intergranular SCC may exist
Net protective Current	• Application limited to bare or ineffectively coated pipe • The measurements are taken at frequent intervals along the pipeline to establish the collection of net cathodic current along the structure	• Bare or ineffectively coated pipe • Not a standard criterion	• Criterion assumes any magnitude of net current flow to the structure (i.e., any amount of CP) is adequate to mitigate corrosion. Because that is not the case, criterion should be considered a last resort • Readings particularly susceptible to misinterpretation due to stray current activity • High-resistivity soils, for deeply buried pipelines, or where the separation distance of the corrosion cells is small also problematic

cathodic reaction: A reaction at the cathode where electrons are transferred to the metal electrode. In pipeline systems, the most common reactions are as follows:

$$2H^+ + 2e^- \rightarrow H_2$$

and

$$O_2 + 2H_2O + 4e^- \rightarrow 4OH^-$$

In these two reactions the anodic reaction is the source of the electrons, which then travel to the cathode and participate in the cathodic reaction. A cathodic reaction is a reduction process.

cation: A positively charged ion (i.e., atom or molecule). In an electrochemical reaction, cations exist in solution and are also produced at the anode. These positively charged ions migrate through the electrolyte from the anode towards the cathode. The migration of cations is illustrated in Figure 34. The movement of anions and cations is fundamental to the effectiveness of cathodic protection systems (See *anode*).

cavitation: The rapid formation and collapse of vapour bubbles in a piping system (such as a pump inlet) whenever the local absolute pressure of a liquid falls below its vapor pressure. The phenomenon can cause damage to adjacent metal surfaces. Cavitation erosion is the progressive loss from the surface of a material due to cavitation.

CD: See *crack detection*.

CDF: See *cumulative distribution function*.

cell: See *electrochemical cell*.

CEPA: See the *Canadian Energy Pipeline Association*.

certainty: A non-standard statistical term used in this context to state the accuracy of in-line inspection (in-line inspection) systems. API 1163RP (recommended practice) states that the accuracy of the tool is to be given as a tolerance, certainty and confidence level. For example, a commonly quoted in-line inspection tool accuracy is ±10% of nominal wall thickness (NWT), 80% of the time. In this stated accuracy, the tolerance is 10%NWT and the certainty is 80%. The statement is equivalent to the statement that the true depth is within ±10% of nominal wall thickness, with an 80% confidence level. However in API1163, the term "confidence level" is reserved for other purposes.

chainage: A linear distance between two points along the pipeline, often measured by means of a chain to accurately record the actual length and elevation due to profile change. Typically, chainage is taken from a fixed reference, such as a compressor/pump station, valve/ or pig launcher. Used interchangeably with the term *kilometer post* or *mile post*.

Survey chainage is usually written with a plus sign "+."In Imperial and US measurement, the distance is measured in feet with the number to the right of the plus sign being in feet and the number to the left of the plus sign is in hundreds of feet. In metric measurements, the number to the right of the plus sign being in meters and the number to the left of the plus sign is in thousands of meters. The number of digits to the right of the plus indicates whether the chainage is in Imperial/US units or metric units.

chalking: Refers to the development of loose, removable powder (pigment) at the surface of an organic coating, usually caused by weathering.

characterization: The process of identifying the type and size of feature detected by an in-line inspection tool. See *defect classification*.

Charpy impact test: Also known as the Charpy v-notch test, this is a standardized high strain-rate test which determines the amount of energy absorbed by a material during fracture in order to measure the material's toughness. The test was developed in 1905 by the French scientist Georges Charpy.

Charpy toughness: Refers to the energy required to break a Charpy sample based on a standardized procedure and sample configuration. Pipelines made of steel with low Charpy toughness are more prone to rupture from corrosion and to cracking. Generally, older pipelines in North America have lower toughness levels (i.e., >100 foot-pounds) – modern pipelines will have toughness values in excess of 150 foot- pounds (i.e., 200 J). A full-sized Charpy specimen has a 10 x 10mm cross-section with a 2mm V-shaped groove machined on one surface. Other specimen sizes are used, and care must be taken to associate the energy value measurement with the type of specimen; e.g., a specimen measuring 10mm x 666mm is referred to as a $CV_{2/3}$.

check valve: A type of safety control valve that is designed to allow the fluid to flow in a given direction but automatically close to prevent flow in the reverse direction. They are also referred to as non-return valves. Check valves can cause damage to in-line inspection tools if they are not fully open as the pig passes; it should be noted that some subtypes of check valves are not piggable at all. There are several types of check valves, including swing check, tilting disc check, and wafer check. Figure 19 shows the basic form of a check valve, and Table 12 lists the major subtypes.

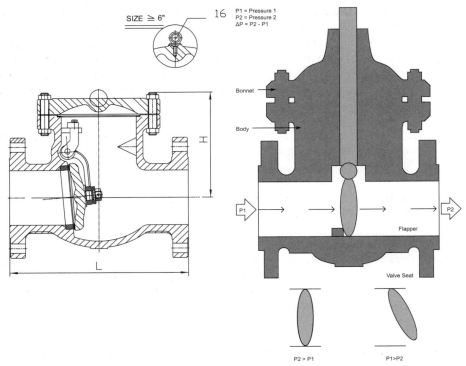

Figure 19. Swinging check valve schematic (a) and operation (b). (200)

Table 12. Major subtypes of check valves (27)

Subtype	Characteristics
Clapper	• The simplest form of the check valve where the internal configuration is such that the valve is not full bore and unlikely to be piggable.
Thru-conduit clapper	• Similar to the clapper style (Figure 19 shows a highly simplified version); however, the valve internals are configured to allow a full bore opening – this style is most likely to be piggable.
Lift	• A valve configuration where the disk and valve seats are horizontally aligned, requiring a significantly different internal valve configuration that redirects flow through a horizontal opening. This style is not piggable.

checking: The development of slight breaks in a coating, which do not penetrate to the underlying surface.

Chi-squared distribution: This is a special case of the gamma distribution, where $\alpha = v/2$ and $\beta = 1/2$. Mathematically, the probability density function of the chi-squared distribution, $\chi^2(x)$, is defined as:

$$\chi^2(x) = \begin{cases} \dfrac{2^{-2v/2}}{\Gamma(v/2)2^\alpha} x^{v/2-1} e^{-x/2}, & 0 < x < \infty \\ \\ 0, & x \le 0 \end{cases}$$

where v = parameter of the distribution, or degrees of freedom

$\Gamma(\alpha)$ = gamma function as follows:

$$\Gamma(\alpha) = \int_0^\infty \left(\frac{x}{\beta}\right)^{\alpha-1} e^{-x/\beta} \frac{1}{\beta} dx$$

Table 13 shows selected properties of the chi-squared distribution.

Table 13. Chi-squared distribution properties

Parameters	v
Domain	$0 \le x \le \infty$
Mean	v
Variance	$2v$
Mode	

Figure 20 shows the shape of the distribution for various degrees of freedom.

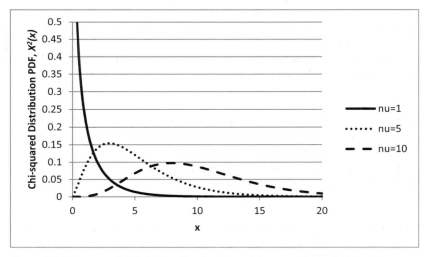

Figure 20. Chi-squared distribution for a variety of degrees of freedom.

The chi-squared distribution is most often used in the chi-squared test which tests the goodness of fit of data to a distribution. The chi-squared distribution is also used to establish confidence intervals, which can be important in risk and reliability assessments. See also *gamma distribution* and *probability distribution*.

chicken scratch: A term used to describe shallow non-significant *stress corrosion cracking* (SCC) on the external surface of a pipeline.

chloride stress corrosion cracking: Cracking of a metal under the combined action of tensile stress and corrosion in the presence of chlorides (compounds containing chlorine) and an electrolyte (usually water).

chronic hazard: A hazard that has the potential to cause long-term damage to health, often as a consequence of repeated or prolonged exposure.

circumferential grooving: Refers to the approximate shape of long narrow metal loss anomalies, circumferentially aligned, as identified through in-line inspection. Specifically, the dimensions of the anomaly are

$$length < width \, / \, 2$$

and

$$A \leq length \leq 3A$$

where A = wall thickness (for wall thickness < 10mm, else $A = 10mm$).

The definition is based on a document published by the *Pipeline Operators Forum*; additional detail and an illustration appear under *defect classification*.

circumferential magnetic flux leakage: An in-line inspection technology similar to conventional magnetic flux leakage (MFL), except that the magnetic field is induced in the pipe wall in the circumferential direction (see *cylindrical coordinate system*) as opposed to the axial direction used in conventional MFL. As a result of the change in orientation, the technology has a much greater resolution in the circumferential direction, but has poorer resolution in the axial direction than conventional MFL. The primary applications to date have been to detect cracking and narrow axially aligned corrosion.

circumferential SCC: This is stress corrosion cracking (SCC) that is aligned with the circumferential direction, around the diameter, of the pipeline. The orientation of SCC is believed to be due to stresses on the pipeline other than the hoop stress from the internal pressure in the pipeline. This is in contrast to most *SCC* found on pipelines, which tends to be aligned to the axial direction.

circumferential sensor: A sensor on a magnetic flux leakage inspection tool that measures the *magnetic flux* (or change in *magnetic flux*) in the circumferential direction – that is, along the θ-direction ("around" the diameter of the pipeline). See *cylindrical coordinate system*.

circumferential slotting: Refers to the approximate shape of long narrow metal loss anomalies, circumferentially aligned, as identified through in-line inspection. Specifically, the dimensions of the anomaly are

$$width > A$$

and

$$length < A$$

where A = wall thickness if $t < 10$, else $A = 10$

The definition is based on a document published by the *Pipeline Operators Forum*; additional detail is provided in the figure under *defect classification*.

CIS: See *close interval survey*.

class location: This term refers to the categorization system of the area around an onshore pipeline based loosely on the consequence of a failure (specifically human life). The population density is used as a proxy for risk and, accordingly, there are four classes, based on the number of buildings and other structures that exist in an area 220 yards (201.2 m) on either side of the pipeline and 1 mile (1.609 km) long. Table 14. Class locations per CFR 192.5 (13) summarizes the population density parameters of the four class locations found in the US regulations under CFR 192.5. Other jurisdictions may have slightly different or numbers of class locations.

classical SCC: See *intergranular SCC and SCC*.

clay: A common constituent of soil. Clay is composed of clastic particles (i.e., previously existing rock that has broken down) less than 0.002 mm (0.077 mils) in size. Clay is commonly deposited in still water such as in lakes (lacustrine), rivers (fluvial), and deltas. Of particular concern for pipelines is that some clay minerals swell when wet. The swelling and shrinkage of the clay puts additional stresses on pipeline coatings, which can lead to disbondment. Tape coatings are particularly susceptible to this stress.

Table 14. Class locations per CFR 192.5 (13)

Location Type	Description
Class 1	• ≤10 buildings / dwellings
Class 2	• 46 < buildings / dwellings AND • >10 buildings / dwellings
Class 3	• >46 buildings / dwellings OR • Pipeline is within 100 yards (91m)of a building • Pipeline is within 100 yards (91m) of a well-defined "outside area" such as playground, recreational area or outdoor theatre • Occupied by at 20 persons 5 days per week for 10 weeks per year*
Class 4	• Locations where buildings with four or more stories are prevalent

*Under Canadian regulations this is a Class 4 location

cleaning pig: A utility pig that uses cups, scrapers, or brushes to remove dirt, rust, mill scale, and other debris from the pipeline. See *pig*.

clock position: See *o'clock position*.

close metal objects: The term used to describe detectable ferromagnetic objects, located close to the external surface of a pipeline, in magnetic flux leakage inspection reports.

close-interval survey: An over-the-line potential survey of a pipeline to assess the effectiveness of the cathodic protection. The survey measures the electrical potential between the pipeline and the soil at regular and closely spaced intervals (usually 1–10 m) along the length of the pipeline. On pipelines with cathodic protection, the survey usually measures both the on potential and off potential. These measurements are then compared to the 850 mV or the 100 mV polarization criteria.

cluster: A grouping of "boxed" metal-loss anomalies, usually as identified by an in-line inspection tool, which are sufficiently close together to be treated as a single unit to calculate burst pressure. Clusters are identified by the application of an interaction rule.

CO_2 corrosion: Refers to a range of corrosion mechanisms that can occur when *carbon dioxide* is present in significant amounts in a pipeline system. While corrosion rates in the presence of CO_2, particularly in combination with water, can reach rates as high as thousands of mils per year – they can effectively be inhibited with chemicals suited to the specific situation (see *inhibitor*).

CO_2 corrosion products include iron carbonate (siderite, $FeCO_3$), iron oxide, and magnetite and can also cause other issues when pigging the pipeline (see *black powder*). Corrosion product colors may be green, tan, or brown to black. CO_2 corrosion is a complex area and the reader is referred to Roberge for a more detailed treatment of the topic; some of the key factors affecting corrosion mechanisms in the presence of carbon dioxide appear in Table 15.

Table 15. Selected factors affecting corrosion in the presence of carbon dioxide (195)

Factor	Description
Water	The presence of water results in the creation of carbonic acid which will attack pipeline steels; however, observed corrosion rates indicate that other mechanisms (such as cathodic depolarization) may also be involved.
Velocity effects	Turbulent flow can often push systems into a corrosive regime by removing and/or preventing the formation of protective iron carbonate (siderite) scale.
Temperature	Higher temperature favors the formation of protective iron carbonate (siderite) scale.
pH	Higher pH (i.e., presence of carbonate waters) favors the formation of protective iron carbonate (siderite) scale.

coal tar: See *coating* (specifically Table 16).

coal tar enamel: Refers to a range of coal-tar based (i.e., derived from the destructive distillation of coal) external pipeline coating formulations that dry to form a hardened physical barrier on the surface of the pipe. See *coating*.

coating: The term refers to the protective layer applied outside or inside the surface of the pipe to act as a physical barrier between the metallic pipe and the external environment or internal fluid to reduce the potential for corrosion/damage or assist the fluid flow internally by reducing pipe roughness. Effectively all modern pipe is installed with an external coating, whereas internal coatings are less common – most often used as a mechanism to increase flow efficiency. Good pipeline coating is characterized by good adhesion and continuity along the pipe surface as well as low diffusion rates for oxygen and water.

The classification of coatings in the pipeline sector is less than uniform and is often referred to by its main constituent (coal tar, asphalt, polyethylene etc.), its form (enamel, mastic, tape etc.), method of application (e.g., field applied, mill applied), or some combination thereof. As such, for the purposes of discussion here, coatings are classified by their main constituent with associated discussion as an attempt to capture other descriptive aspects consistently. Figure 21 shows the time periods during which various coatings have been used. Table 16 lists the various coating types and their properties. Table 17 lists the properties of coatings. It should be noted that the following discussion refers primarily to pipe body coatings – girth weld coatings are typically all field applied and vary somewhat from the information provided here.

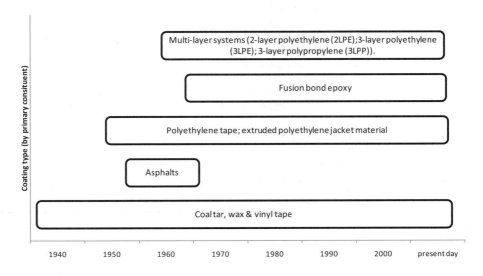

Figure 21. Approximate timeframes for usage of various coating types. (181)

Table 16. Selected considerations in selecting a pipeline coating (168)

Main constituent	Description	Strengths	Weaknesses
Asphalts	• Derived from fractional distillation of bitumen • Introduced more than 75 years ago this bituminous mixture is usually applied in a thick layer (1/2 to 5/8 in.) • The mixture often contains select graded sand, crushed limestone, and glass fiber • Seen in both enamel and mastic forms; less commonly in tape form • Application of mastic/enamel is primarily through extrusion; at times used with various outer wraps (similar to coal tar)	• Minimal susceptibility to holidays • Low current requirements • Good resistance to cathodic disbondment • Good adhesion to steel	• Handling limitations (heavy and bulky) • Limited availability and often costly • Susceptible to high temperatures degradation mechanisms
Coal-tar	• Derived from destructive distillation of coal • Usually used in enamel form with an outer wrap acting as a backfill shield. Outer wraps were initially rags before the use of asbestos felt; resin-bonded glass fiber mats are now used since asbestos has been discontinued • Coal tar enamels are still used internationally but use in North America is almost nonexistent now • Modern enamels allow operating temperatures to 230°F. • Some epoxy mixed formulations available (see below)	• Minimal susceptibility to holidays • Low current requirements • Good resistance to cathodic disbondment • Good adhesion to steel	• Limited manufacturers and applicators • Health and air quality concerns • Older coatings do not tolerate high temperatures well and are especially susceptible to creep and cold flow due to soil stress

continued next page

Table 16 *continued*

Main constituent	Description	Strengths	Weaknesses
Epoxy	• Based on resins formed by the reaction of aliphatic or aromatic polyols. • Initially only available for NPS 8 pipe and smaller • Early formulations required a primer • Most formulations largely unchanged in last ~20 years • Can be either field or plant applied • Most common form, fusion bond epoxy (FBE), involves spraying a powder formulation on the heated pipe surface • Some liquid forms based on a coal tar formulation (i.e., coal tar epoxy)	• Good resistance to soil stress and cathodic disbondment • Low current requirement • Excellent adhesion to steel • Excellent resistance to hydrocarbons	• Exacting surface preparation required • High application temperature • Low impact resistance • High absorption of moisture
(Mixed) Multi-layer systems	• Multilayer systems are an approach to use the best characteristics of several different constituents for superior performance – inner layers tend to be FBE (or similar) with the outer polyethylene/polypropylene layers providing impact and abrasion resistance • The most common systems are: ◊ 2-layer polyethylene (2LPE); ◊ 3-layer polyethylene (3LPE); ◊ 3-layer polypropylene (3LPP).	• Lowest current requirements • Highest resistance to CP disbondment • Excellent adhesion to steel • Excellent resistance to hydrocarbons • High impact and abrasion resistance	• Limited applicators • Exacting application parameters • Higher initial costs • Possible shielding of CP current

continued next page

Table 16 *continued*

Main constituent	Description	Strengths	Weaknesses
Polyethylene	• Fabric-reinforced petrolatum-coated tapes first used ~65 years ago with the more common field applied versions introduced ~50 years ago; mill-applied tape systems introduced ~30 years ago. • Mill applied systems are actually multi-layered (a primer, a corrosion-preventative inner layer of tape, and one or two outer layers for mechanical protection) • CP shielding concerns led to development of fused multi-layer tape systems with an electrically conductive backing • Extruded coatings polyethylene coatings can be a very dependable and cost effective solution – most often used for smaller diameter pipe (often referred to as Yellow JacketÒ, which has become generic name for coating class)	• Tapes • Ease of application • Low energy for application • Extruded polyethylenes • Good handling characteristics • Good protection from mechanical damage	• Tapes • Handling restrictions (shipping and installation) • UV and thermal blistering • CP shielding • Susceptible to soil stress disbondment • Extruded polyethyelenes • CP shielding a concern in disbonded areas
Vinyl	• Primarily used in field or mill applied tape form	• Ease of application • Low energy for application	• Handling restrictions (shipping and installation) • UV and thermal blistering • CP shielding • Susceptible to soil stress disbondment
Wax	• Primarily used in field or mill applied tape form (i.e., hot applied) • Used both with and without outer wrap/ backing	• Ease of application	• Relatively low CP shielding

Table 17. Summary of coating types, strengths and weaknesses (168)

Property	Description
Abrasion resistance	Especially important during the transportation of the pipe and construction of the pipeline.
Adhesion	The ability of the coating to remain on the pipeline and to prevent water from seeping under the coating
Cohesive strength	The resistance of the coating to break or crack.
Dielectric strength	High electrical resistance is important because corrosion is an electrochemical reaction and to reduce the load on the CP system; however, when the coating does fail, it is ideal that it fails in a manner that does not shield CP (this is more of a mechanical characteristic rather than a material property).
Ease of repair	Field repair of the coating at some locations is inevitable. The repaired coating should have properties equal to or better than the original coating.
Flexibility/ elongation	The coating must be able to bend with the pipeline without cracking or disbanding; however, it must be resistance to sagging/ disbondment resulting from soil stresses
Holiday resistance	The coating should cover the pipe surface complete with no holes and resist the development of holidays over the life of the coating.
Impact resistance	Especially important during the transportation of the pipe and construction of the pipeline.
Impermeable	The coating must prevent water and other corrosive agents from reaching the surface of the pipe to avoid corrosion.
Toxicity	Pipelines coating must be nontoxic to the environment.
Temperature resistance	The coating must retain its physical and chemical properties over the full range of temperatures it will experience.
UV resistance	The coating must be resistant to UV radiation during storage and stockpiling.

coating system: Refers to all of the specific components needed to ensure that a pipeline coating is applied in the manner intended by the manufacturer. The term is most often used where multiple layers or components are used. See *coating*.

code: Refers to a procedure for the design, construction, or maintenance of an engineering structure as provided in a standard adopted by the regulating body that has jurisdiction as being a requirement or test of acceptable practices.

coil sensor: Used to measure the rate of change in magnetic flux – typically on an in-line inspection tool. Coil sensors are older technology where a coil of wire is used to measure a change in total magnetic flux. That is, a change in the total magnetic flux, through the center of the coil, induces a current in the coil according to Faraday's Law of Induction. Modern pigs use *hall-effect sensors* because of their smaller size and because they enable the measurement of the actual magnetic flux rather than the rate of change of the field.

cold applied coating: Refers to the range of external pipeline coating formulations that do not require the use of heat for effective application – regardless of the constituent compounds. Thus, this is a generic term often used to describe the numerous coating systems most often used in the field for repairs, girth welds and tie-ins, where there is no mill-applied coating. See *coating*.

cold bend: Refers to pipe bending that is done in the field – as such, no heat treatment is applied. Due to the potential for strain hardening (and in the extreme, pipe buckling) and coating damage, there are limitations on the amount of curvature that can be introduced in cold bending, most often 1.5 degrees per diameter The total required bend angle is created by a series of narrowly spaced small bends. The minimum radius of curvature allowed during cold bending is a function of the pipe diameter, whether the pipe is coated, the amount of wall thinning, and the location of girth welds as defined by the codes and standards relevant to the jurisdiction.

cold bend crack: Damage that occurs to the external coating of a pipe joint during *cold bend* operations because the flexibility of coating is reduced at low temperatures. The cracks tend to have a circumferential orientation.

cold bend kink: Also called a ripple, a cold bend kink can result along the circumferential direction of the pipeline if a cold bend is performed incorrectly, most often by too great a pull. Usually referred to as ripples on the exterior curvature, wrinkles on the interior curvature. See *cold bend*.

cold cracking: See *Hydrogen cracking.*

cold work: Refers to plastic deformation of a metal at a temperature below the recrystallization temperature and at a strain rate that induces strain hardening. Usually, (intentional) cold working is done at room temperature. In some cases, cold working is done to achieve a desired hardness of the metal. In terms of pipeline integrity, the more relevant discussion of cold work is in the context of highly localized areas that are effectively "cold worked" as a result of damage (such as denting, gouging, etc.). Specifically, these regions of cold work have lower ductility and toughness and are therefore more susceptible to several forms of cracking (e.g., fatigue and environmentally assisted cracking).

colony: A grouping of cracks such that they are confined to a relatively small area and are potentially interacting from the point of view of pipe stresses. The number of cracks in a colony can range from just a few to thousands. The term is applicable to all types of cracking but is most often used in the context of stress corrosion cracking.

commercial blast cleaned surface: A term used to describe the level of cleaning to a metal surface. A commercial blast cleaned surface, when viewed without magnification, is free of all visible oil, grease, dust, dirt, mill scale, rust, coating, oxides, corrosion products, and other foreign matter. Random staining is limited to 33% of the surface and may consist of light shadows, slight streaks, or minor discolorations caused by stains of rust, stains of mill scale, or stains of previously applied coating.

composite repair: any of a range of repair techniques using a synergistic combination of two or more materials and placed over the damaged area in the form of a sleeve or an epoxy-filled shell. Composites comprise high-strength reinforcement in fibrous form, incorporated into and bonded together by a matrix, usually a thermosetting polymer. The most common strength fibrous component used is glass.

composite sleeve: A sleeve constructed of a composite material. See *sleeve*.

compression sleeve: A sleeve used to repair some type of pipeline anomalies. See *sleeve*.

compressional wave: Also called longitudinal waves, these are acoustic body waves (a subset of elastic waves) that are often used in ultrasonic testing of pipelines. The waves are characterized by atoms and molecules in the bulk material moving in a direction in-line with the direction of the wave propagation. Unlike other elastic waves, they will travel in gases and liquids; however because of the contrast in acoustic impedance, it is very difficult to transmit acoustic energy from the gas to steel or steel to gas. See *ultrasonic testing* for a more detailed discussion of the topic.

compressor station: An installation along a gas pipeline used to facilitate flow of gas in the pipeline by pressurizing the product. Typical compressor stations consist of the compressor, associated valves, piping, station scrubber, and may also include a gas cooler to allow higher compression ratios and/or to meet coating temperature limits and reduce pressure-drop along the pipeline.

concave cap/root: See *excess concavity*.

concentration cell: See *differential aeration cell*.

concentration controlled process: A corrosion reaction in which the corrosion rate is limited by the rate at which hydroxyl ions (for example) can diffuse to the surface of the metal to receive an electron and complete the circuit.

concentration polarization: The portion of total polarization of a cell caused by the limited diffusion rate in the electrolyte. Thus, the term is relevant for interpreting and applying cathodic protection criteria.

The total polarization of a corrosion reaction is given by

$$E - E_0 = \eta_{tot} = \eta_{act} + \eta_{conc} + iR$$

where
E = potential of the corrosion cell
E_o = equilibrium potential
η_{tot} = total electrochemical polarization
η_{act} = total activation polarization
η_{conc} = total concentration polarization
iR = ohmic drop due to the flow of current through a resistive medium

The concentration polarization is a function of the diffusion rate in the electrolyte, and is given by

$$\eta_{conc} = \frac{2.3RT}{nF} \log\left(1 - \frac{i_c}{i_L}\right)$$

where
$$i_L = \frac{D_z nFC_n}{\delta}$$
R, T, n, F = thermodynamic constants and parameters
i_L = maximum current density through the electrolyte
C_n = species concentration
δ = constrictivity constant
D_z = coefficient of diffusion

Figure 22 shows the concentration polarization as a function of the current density. The finite diffusion rate limits the maximum current possible. The concentration polarization is the result of that limit.

condition monitoring: Refers to the numerous methods to monitor and assess the state of the pipeline. See *pipeline monitoring methods*.

conductivity: 1. A material property relating heat flux (heat transferred per unit area per unit time) to a temperature difference. **2.** The ability of a water sample to transmit electric current under a set of standard conditions. Conductivity is the inverse of resistivity and it is usually expressed in mho (ohm^{-1}).

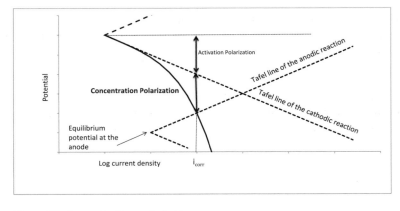

Figure 22. Concentration polarization.

confidence interval: The interval around which the true value of some parameter is expected to be, at a specified frequency. For example, if the depth of a corrosion anomaly is reported to be 3 mm ± 0.5 mm, 95% of the time, then the confidence interval is 2.5-3.5 mm. Formally, a confidence interval is written as a set; for example, if value of α is between a and b, then

$$\alpha \in (a, b)$$

However, it is common to state the interval in terms of a tolerance:

$$\alpha = \frac{a+b}{2} \pm \frac{b-a}{2}$$

The size of a confidence interval is dependent on the confidence level. Figure 23 illustrates the confidence interval.

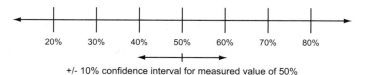

+/- 10% confidence interval for measured value of 50%

Confidence interval

Figure 23. Confidence interval.

confidence level: Used in conjunction with a confidence interval to state the probability or proportion of instances for which the true value of some parameter is in the confidence interval. The confidence level is usually stated explicitly. If the depth of a corrosion anomaly is reported to be 3 mm ± 0.5 mm, 95% of the time, then the confidence level is 95%. The most common use of this term is in stating a confidence level for in-line inspection reported depth, length, or width, a confidence level of 80% is common.

consequence: This term, which can be defined by a range of scenarios, is one of the key components of a risk analysis. It is an expected outcome, or societal impact, that would result should a specific event occur. Often, consequence is quantified as a dollar value or commercial impact, but other units are possible. Specifically, it is intended to measure the damage or injury to human life, property, and the environment due to a specific event (such as the release of a hazardous material from a pipeline). The way in which consequence is calculated depends on the type of risk being calculated: quantitative, perceptional, individual, societal, or territorial.

consequence matrix: See *qualitative risk.*

construction defect: A pipeline anomaly or defect that is introduced during field construction. About 5% of pipeline failures still result from some form of damage resulting from past construction practices. Most of these anomalies are related to the completion of the girth weld (see *weld defects*), but also include non-weld-related anomalies. The nature of these defects is broad ranging, and, as such, the tools and techniques to determine whether they are injurious also vary significantly. Table 18. Construction defects and related integrity concerns [adapted from (14)] lists examples of non-weld-related construction defects.

Table 18. Construction defects and related integrity concerns [adapted from (14)]

Defect	Description
Butt welded patch	A technique seen in older pipelines where a patch is applied to the pipe body to remedy various situations, including repair of an extensively damaged area of pipe or to cover a manhole. The integrity of the weld, specifically the material properties, can pose an integrity concern.
Cold bend kink	A situation where the pipe body is significantly deformed as a result of the field bending process. Localized strain hardening and the potential for cracking in the area of the kink can pose an integrity concern.
Close metal object / eccentric casing	A situation where a metal object is in direct contact with the external pipe wall – this potentially has significant implications for the cathodic protection system on the pipeline. Eccentric casings in particular pose a problem because they have sufficient mass to effectively short circuit the cathodic protection system.
Lack of cover	A situation where the depth of backfill over the pipeline is less than specified. Insufficient backfill can increase the potential for issues including, but not limited to, mechanical damage, unacceptable deformations due to ground movement, or the need for buoyancy control.
Mitre bend	Technically speaking mitre bends (i.e., the use of angle cuts on the pipe to facilitate a change in direction instead of using a fitting or pipe bending) are intentional and therefore not a defect in the pipe. Especially since the use of mitre bends does not directly pose an integrity concern. However, mitre bends can result in pig damage during inspection operations and may pose a threat in that sense.
No field joint coating	Poor quality control may result in individual girth welds not being coated (i.e., if the pipe was welded in the field) – potentially resulting in a greater possibility of all forms of environmentally assisted damage (i.e., cracking and corrosion).

contact corrosion: See *galvanic corrosion.*

continuity bond: Refers to any form of electrical contact between a pipeline and a foreign structure. In extreme cases, the continuity bond disrupts the cathodic protection system to the point that the pipeline system is left entirely unprotected.

copper-copper sulfate electrode: This is one of the most commonly used *reference electrodes* when measuring various potential levels for testing the cathodic protection levels of a buried pipeline. The electrode is constructed with a solid copper bar immersed in a saturated copper sulfate solution. Its advantage over other electrodes is that it is simple and rugged for field operations.

corrosion: Defined generally as the deterioration of a material that results from a reaction with its environment. Usually, corrosion refers to the oxidation of a metal in an electrochemical reaction. Most metals are unstable in the presence of water and undergo oxidation. Corrosion reactions require an anode site, a cathode site, an electrolyte, and an electrical pathway.

For steel pipelines the corrosion reaction of interest is that of iron. One of the reactions by which iron corrodes is

$$Fe + 2H_2O \rightarrow Fe(OH)_2 + H_2$$

Corrosion reactions can be separated into two half-cell reactions: oxidation and reduction. The corrosion of iron can be separated into the oxidation reaction:

$$Fe \rightarrow Fe^{2+} + 2e^-$$

and the reduction reaction:

$$2H_2O + 2e^- \rightarrow 2OH^- + H_2$$

corrosion coupon: The use of corrosion coupons is one of the simplest methods of monitoring corrosion activity. A coupon is a small sample of material representative of the pipeline. The coupon is placed at a location in or near the pipeline where corrosion is suspected. After a specified period of time, the coupon is removed and examined for corrosion. While corrosion coupons provide a relatively realistic assessment of corrosion in the context of specific site conditions, relatively long timeframes are needed to obtain meaningful data.

corrosion deposit: Refers to the accumulation of a range of chemical products in the area of the pipe where corrosion is or has been active.

corrosion fatigue: A cracking mechanism encountered on pipeline systems – although with a frequency significantly lower than other mechanisms, such as general corrosion or stress corrosion cracking. Specifically, corrosion fatigue occurs when repeated cyclic loading and corrosive conditions exist simultaneously. The final result is such that the component experiences failure at lower stress levels, or fewer cycles, than would have occurred in the absence of the corrosive environment.

corrosion growth analysis: The term typically used for a study focused on determining the rate at which corrosion is growing. The study is most often conducted when an operator feels that "generic" industry standard assumptions are not sufficiently accurate for the specific situation that is being assessed. In the context of pipeline integrity, the most common forms of study are shown in Table 19.

Table 19. Selected approaches to establishing corrosion growth rate

Technique	Description
Physical testing	Techniques range from lab based studies of corrosion rates to the use of *corrosion coupons*.
In-line inspection data	The primary approach is to use data from successive in-line inspections. The reported metal loss anomalies from each inspection are correlated, and their size (depth, length, and width) or signals (magnetic or ultrasonic – based on the technology used) are compared. Any increase in the metal-loss depth may indicate active corrosion – especially where this increase in depth is well beyond the error bounds of the inspection tools and interpretation methods. Usually, these methods heavily rely on statistical methods.

corrosion growth rate: See *corrosion rate*.

corrosion inhibitor: See *inhibitor*.

corrosion mapping: The process of documenting areas of general corrosion, usually in the context of an excavation, where depth measurements (or remaining wall thickness measurements) are taken using a finely spaced *grid*. Corrosion mapping is usually displayed as a profile, grid or a colored contour map and the data are typically used as input to burst pressure calculations. The main technologies available for corrosion mapping are shown in Table 20.

corrosion monitoring: The process of monitoring the processes and rates of corrosion either continuously or by periodic inspection. There are numerous methods of monitoring corrosion thatcan be classified as direct or indirect methods. In the context of pipeline integrity, some of the most common methods include the use of *smart pigs* as well as *corrosion coupons*.

Table 20. Main corrosion mapping techniques [adapted from (15)]

Technique	System Type	Description
Rubbings	Manual	• The process uses graphite literally "rubbed" over a piece of paper placed over the corroded area • Deeper areas appear as more darkly shaded regions • Typically used for deepest areas of corrosion • Limited usefulness in conducted quantified analysis (i.e., burst pressure calculations)
Pit gauge readings	Manual	• The use of a *pit gauge* (mechanical instrument) to establish corrosion depth readings usually on a grid basis (grid spacing typically ranges from 0.25 to 1 in.) • A key source of error is if the probe is relatively large compared to the diameter of the corrosion spot measured
Pencil probe readings	Manual / automated	• The use of a *pencil probe* (usually an ultrasonic based instrument) to establish corrosion depth readings usually on a grid basis (grid spacing typically ranges from 0.25 to 1 in.) • Ultrasonic technology based systems are also available mounted on assembly that facilitate "automated" data collection
Laser mapping	Automated	• The use of laser-based instrumentation (almost always mounted on an assembly) to establish a corrosion depth reading basis (grid spacing is typically finer than the spacing that is practical for pit gauge or pencil probe readings) • Highly sensitive to surface preparation • See also *laser mapping*
In-line inspection tools	Automated	• A range of technologies (*MFL, ultrasonic,* etc.,), mounted on a pig, are used to map corrosion on the internal and external surfaces of the pipeline

corrosion potential: The electrical potential of the surface that is corroding when compared to a reference electrode when the cathodic protection system is turned off (i.e., under open-circuit conditions). This term is used interchangeably with *rest potential*, *open-circuit potential*, or *free corrosion potential*.

corrosion product: A substance formed as the result of a corrosion reaction. See *corrosion deposit*.

corrosion rate: The rate at which corrosion proceeds, or grows, per unit time. The rate is typically measured as either the rate of depth increase or as a weight loss rate. Change in depth is usually stated in units of millimeters per year (mm/year) or inches per year (in/year) (i.e., *corrosion growth rate*). Weight loss is usually stated in units of grams per square meter per year ($g/m^2/year$) or ounces per square foot per year ($oz/ft^2/year$).

In the context of pipeline integrity, the *corrosion growth rate* is the more commonly used parameter in maintenance planning. Specifically, the expected corrosion growth rate is often used to establish in-line inspection frequencies and establish multi-year excavation/repair plans (based on the expected time to failure of known anomalies).

corrosivity: The tendency of an environment to cause corrosion in a given corrosion system. In the context of pipeline integrity, the corrosivity of the "environment" is ultimately determined by a combination of factors including, but not limited to, soil properties (pH, resistivity, chemical composition, homogeneity), and the presence of moisture (amount of water present, composition etc.,). No overarching method for determining corrosivity exists – rather a number of *soils models* exist in the industry specific to threats such as *stress corrosion cracking*.

couplant: A medium used to in the *ultrasonic* inspection of a solid material such as a pipe wall. The purpose of a couplant is to increase the transmission coefficient of ultrasonic energy into the material being inspected. Without a couplant, ultrasonic energy would need to be transmitted through air or gas before being transmitted into the material (i.e., air/gas will exist even in a small gap between the pipe wall and a transducer). Because materials like steel have an acoustic impedance that is very different than air, almost no energy is transmitted from an ultrasonic probe into the steel. The couplant's acoustic impedance is more similar to steel, which allows sufficient energy to be transmitted into the steel to perform the inspection.

NDT personnel use a range of commercially available couplants when performing an ultrasonic inspection of a pipeline during an excavation. In-line ultrasonic inspections face similar challenges when using *ultrasonic* based techniques from the inside of a natural gas pipeline. If necessary, ultrasonic inspection tools can be used in gas pipelines with a slug of water, diesel, or similar material to enable the transmission of the acoustic or elastic energy into the steel.

couple: See *galvanic couple*.

coupon: See *corrosion coupon.*

CP: See *cathodic protection.*

CP criteria: See *cathodic protection.*

crack: A narrow fissure in the pipe material, usually with uneven surfaces resulting from or through multiple microscopic level brittle fractures. A crack may be *intergranular* or *intragranular*. That is, the growth of *intergranular* cracking mechanisms is influenced by the microscopic material properties and the defect preferentially follows the contours of grain boundaries. In the case of *intragranular* cracks, the direction of crack growth is unaffected by the grain boundaries (although it may be influenced by other microscopic material characteristics, such as highly localized variations in ductility or toughness).

crack arrestor: Any one of several mechanisms used to establish *fracture control* (i.e., reduce the length of any running fracture that may occur on the pipeline). Specifically, crack arrestors work by decreasing the pipe stress either through reinforcement of the pipe wall (a common example is the use of a reinforcement sleeve) or an increase in strength (a common example is the insertion of heavier wall or higher strength pipe joints at regular intervals).

crack coalescence: Refers to the merging of two or more separate *cracks* to form a longer crack. The coalescence of *stress corrosion cracks* is common because of the close proximity of the cracks in a colony.

crack detection: The process of identifying cracks, or crack-like anomalies, in a pipe wall by *non-destructive testing* methods. Common methods to detect cracks in a pipeline are *magnetic particle inspection, eddy current*, and *angle-beam ultrasonics.*

crack detection tool: A *pig* that is designed to detect cracks and crack-like anomalies in a pipeline. Technologies that are used by in-line inspection pigs for the detection of cracks include *ultrasonics* (UT), *electromagnetic acoustic transducer* (EMAT), and *eddy current.*

crack initiation: Refers to the mechanical and chemical group of processes, occurring at the microscopic level, that lead to the development of a crack. It is the first of what can be considered three stages of the life of a crack. See Table 21.

crack propagation: See *crack initiation.*

Table 21. Three stages of crack life (10)

Stage	Description
I - Initiation	• Often considered the longest of the three stages • Very difficult to predict where cracks will initiate
II Propagation	• The second stage that essentially involves the growth of the crack to the point just before failure • This is the most visible / detectable stage (short of failure); thus, techniques to mitigate the threat of cracking tend to focus on this stage – as opposed to the initiation stage
III - Failure	• The relatively short and final stage where the remaining wall thickness of the pipe is insufficient to support the internal stresses

Factors controlling crack initiation include corrosion, cyclic loading, and inclusions.

cracking: This term refers to the formation, or existence, of cracks primarily in the propagation stage. Cracking can be the result of various root causes in pipeline metals, but are usually a result of several parameters, including the microstructure of the metal and the presence of stress; these effects can be exaggerated by the presence of hydrogen, a corrosive environment, cycling of stresses, and defects introduced through the fabrication process. The main types of cracking found on pipeline systems are *fatigue cracking*, *high pH stress corrosion cracking*, *near-neutral stress corrosion cracking*, *hydrogen-induced cracking* and *corrosion fatigue*.

cracking (of coating): Cracking, in external pipe coating, is also important in the context of pipeline integrity. When the term is used in reference to a coating defect, it refers to a complete break in the layer through to the underlying coating layer or pipe wall. There several types of coating cracks – the most common appear in Table 22. Various types of cracking in coating (15).

Table 22. Various types of cracking in coating (15)

Crack type	Coating type	Description
Cold bend cracks	Coal-tar enamel, asphalt, bitumen, and fusion-bond epoxy.	Irregular circumferential cracks at a field bend. Cracks in fusion-bond epoxy can occur at low temperatures and high stress levels.
Hydrostatic test cracks	Coal-tar enamel, asphalt and bitumen	Irregular axial cracks which have resulted from the expansion of the pipe during a hydrostatic test. Cracks in fusion-bond epoxy can occur at low temperatures and high stress levels.
Soil cracks	Coal-tar enamel, asphalt and bitumen	Cracking of the coating due to the loading and movement of the soil around the pipeline.

crawler tool: A self-propelled inspection tool or *pig*. Unlike other inspection tools (sometimes called free-swimming tools), crawler tools are powered by an external source and driven through the pipeline on wheels rather than being pushed along by the pipe product. Crawler tools are often tethered by power cables and communication lines.

The technology allows the inspection of otherwise unpiggable pipelines, by enabling the pig to enter a pipeline at one end, inspect a length of pipeline, and return to the same end for retrieval. The length of pipe that crawler tools can inspect is limited by the length of the cables as well as the number and nature of bends present in the pipeline. Typically, crawler tools can inspect up to about 10 km of straight pipe and is more likely to be required for smaller diameter pipelines.

crazing: This a network of fine cracks, usually interlaced, that appear on the surface of a pipeline coating. Crazing is usually a result of improper construction procedures – including, but not limited to, improper curing, excessive build-up of coating, improper mixing of coating ingredients or environmental conditions (i.e., temperature, humidity, etc.) that are significantly different from those specified for coating application or performance conditions. If the crazing is severe, the protective ability of the pipeline coating may be compromised.

creep: The time-dependent strain that occurs in a component due to stress. When the creep strain occurs at a diminishing rate, it is called primary creep. A minimum, and nearly constant creep rate, is termed secondary creep. Accelerating creep is called tertiary creep. There is a school of thought in industry that creep may well play a role in hydrostatic testing. For example, the British pipeline standard (PD 8010-2) indicates that a test must be held at pressure for at least 24 hours to address the possibility of failure after several hours due to creep.

crevice corrosion: This is the oxidation of a metal surface at, or immediately adjacent to, an area that is sheltered from the ambient environment. Crevice corrosion occurs at a site where disbonded coating or other object restricts the diffusion of oxygen to that location. The difference in oxygen levels creates a differential aeration cell.

critical defect: As defined by several NACE standards, this is a pipeline anomaly (either corrosion, cracking, mechanical damage or other defect) that is predicted to fail under normal pipeline operating pressure and conditions. A critical defect requires immediate attention to avoid failure.

critical flaw size: The size of a flaw in a structure that will cause failure at a particular stress level.

crude oil: One of several unrefined (liquid) hydrocarbon products carried by pipelines. Crude oil is a mixture of hydrocarbon liquids, other liquids, such as water, and gasses, such as natural gas. Chemically, it is composed of thousands of substances – although these are primarily from one of three groups shown in Table 23.

The physical properties of crude oil vary widely with its chemical properties: Its color may be clear, green, amber, brown, or black. Its density is often stated as its *API gravity*. Most crude oil has an API gravity between 10-70. The most valuable crudes have an API gravity of 40-45 (specific gravity of 0.80). Heavy oil has an API gravity less than 22.3. The heaviest crudes can be heavier than water with an API gravity of 8 and can be extremely viscous – often needing heating or solvent mixing for pipeline transport.

Table 23. Key components of unrefined crude oil (10)

Hydrocarbon type	Key Characteristics
Paraffins	Paraffins are a series of straight-chained hydrocarbon compounds. They have chemical formulas of the form $C_n H_{2n+2}$. The simplest paraffin is methane (CH_4), followed by ethane, butane and propane.
Aromatics	Aromatics are a series of hydrocarbon compounds that have a single closed ring structure with single and double bonds. Benzene ($C_6 H_6$) is the most common member of this group. They have chemical formulas of the form $C_n H_{2n-6}$.
Naphthenes	Naphthenes are a series of hydrocarbons with have a molecular structure of a single closed ring and have of only single covalent bonds. They have chemical formulas of the form $C_n H_{2n}$
Asphaltenes	Asphaltenes are hydrocarbons that have a high-boiling point and high-molecular weight and contain many other atoms, such as sulfur, oxygen, nitrogen, and other inert matter.

CSA: See *Canadian Standards Association.*

CSA Z245.1 Steel Pipe: This Canadian standard covers the requirements for steel pipe intended to be used for transporting fluids as specified in CSA Z662, Oil and Gas Pipeline Systems.

CSA Z662 - Oil & Gas Pipeline Systems: Canadian standard for the design, construction, operation and maintenance of primarily of oil and gas pipelines. The standard also covers oil field water, steam, and carbon dioxide.

CSE: See *copper-copper sulfate electrode.*

CTE: See *coal tar enamel.*

Cu/CuSO₄: See *copper-copper sulfate electrode.*

curing: The chemical process of hardening a polymer or resin by cross linking the polymer chains. The hardening process can be initiated by several different mechanisms, such as adding a chemical substance or heating to create a permanent protective coating on the (internal or external) surface of the pipeline. Most modern pipeline coatings require curing to achieve the desired properties.

curing agent: A chemical substance used to initiate the curing process – usually for pipeline coatings. See curing.

current: See *electrical current.*

current density: The amount of current flowing per unit area. In the context of pipeline integrity, and more specifically corrosion, the current density is a key parameter related to the corrosion rate. The standard unit of current density are A/m^2, or mA/cm^2.

current efficiency: The ratio of the electrochemical equivalent current density for a specific reaction to the total applied current density.

current interrupter: A device used to interrupt cathodic protection. The device is useful for a range of surveys that require that measurements be taken with the cathodic protection system being turned off.

cut out: A type of pipeline repair where the part of pipe containing the defect is physically cut out and replaced with a new joint of pipe. In addition to being significant in the context of pipeline repair, a cut out will also result in girth-weld numbering used to report in-line inspection data – thus, changes in girth-weld numbering must be addressed if attempting to compare data between successive inspections. Figure 24 illustrates the impact of the cut out on the reported girth-weld numbering in inspection data.

cylindrical coordinate system: This is a commonly used system to locate and describe items (e.g., the location of anomalies) on pipelines. Figure 25 shows the typical definition of a cylindrical coordinate system.

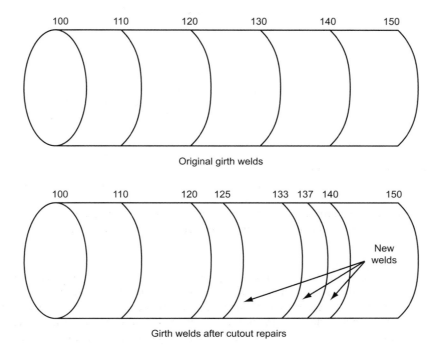

Original girth welds

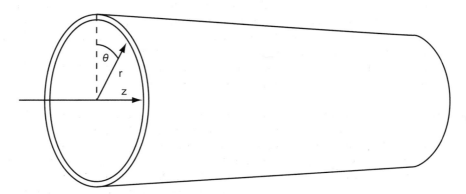

Girth welds after cutout repairs

Figure 24. ILI report girth-weld numbering after a cut-out repair.

Figure 25. Cylindrical coordinate system.

D

D/S: See *downstream*.

DA: See *direct assessment*.

data analysis: The term most often used to describe the process of interpreting and evaluating indications recorded by an in-line inspection to classify, characterize, and size them.

data integrity: Refers to the quality of a given data set; that is, good data integrity implies that the data is current and accurate. Given the nature of pipeline integrity issues, data integrity is an essential component of pipeline integrity: good operational and maintenance related decision making often relies on data intensive inputs, such as results of in-line inspection, cathodic protection surveys, population, and *GIS* location information among others.

data management: Usually, a loosely defined term in the context of pipeline integrity; however, the international Data Management Association (DAMA) defines data management as "….the development and execution of architectures, policies, practices and procedures that properly manage the full data lifecycle needs of an enterprise….". Specifically, the association has defined two frameworks (consisting of 10 data management functions and 7 environmental elements) that require consideration in tackling data management related issues. Figure 26 shows the four elements.

database: Formally, data that is organized and managed in a very specific way allowing the user to query and combine information relatively quickly and efficiently in a range of ways. Databases require specialized software systems called database management systems to organize and maintain the data. Information is retrieved from a database using specialized computer languages such as SQL (Structured Query Language). Informally, the term "database" is often used to describe any collection of data whether organized or not.

day lighting: Excavation or exposure of a pipeline so that a direct physical examination can be conducted.

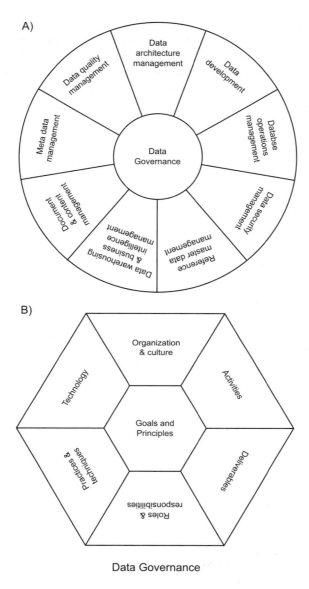

Figure 26. The main elements of data management systems as identified by DAMA. (114)

DC: See *direct current*.

DCVG: See *direct current voltage gradient*.

dead leg: This is a portion of pipeline where the termination point is a dead end and as such there is little circulation and no fluid flow. The section of pipe then becomes an area that traps water, bacteria, and other materials – potentially increasing the risk of internal corrosion. Depending on the geometry of the dead leg, the ability to flush the undesirable substances maybe very limited – if at all possible. A dead leg is illustrated in Figure 27.

dead weight pressure tester: A simple device used to measure pressure – most often in the context of pressure testing of a pipeline. This type of mechanism is relatively accurate due to reliance on the principle of P = F/A, where the pressure to be measured exerts a force, F, on a sealed piston of area A within a cylinder, which is then balanced against previously calibrated weights. Figure 28 illustrates the principle of a dead-weight pressure tester.

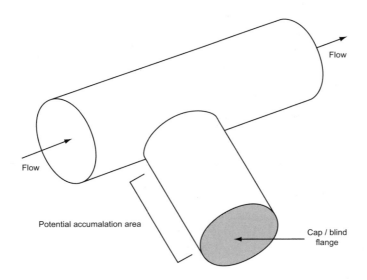

Flow

Flow

Potential accumalation area

Cap / blind flange

Figure 27. Dead leg and potential liquid/debris collection points.

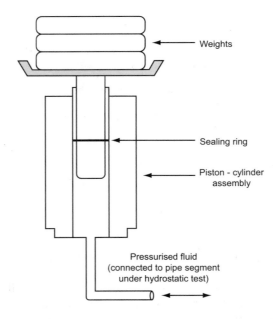

Weights

Sealing ring

Piston - cylinder assembly

Pressurised fluid
(connected to pipe segment
under hydrostatic test)

Figure 28. Principle behind the operation of a dead-weight pressure tester.

deadend: See *dead leg*.

decommission: Decommissioning is the process of taking a facility (or group of facilities) out of service. This term is often used interchangeably with the term *abandonment* but strictly speaking, it is technically one of the steps involved with the *abandonment* process. See *abandonment*.

deep ground bed: A ground bed is the installation of one or more anodes in drilled holes as part of a cathodic protection system; thus, a deep ground bed refers to the installation of anodes at a vertical depth of 15 m (50 ft) or more below the earth's surface.

defect: A term interchangeable with *anomaly*, but in recent years, especially in the United States, defect has come to mean a material flaw that, by its existence, potentially compromises the ability of a component to either perform as per its design basis or meet minimum acceptable standards/specifications. The distinction is not universally observed.

defect assessment: Refers to the analysis and study of a discontinuity in the pipe material to determine whether it is injurious and, if it is, to determine the optimal repair method.

defect classification: Usually referred to the classification of an anomaly based on the results of an in-line inspection. The exact classification nomenclature varies from vendor to vendor; however, the classification can be expected to be similar to that presented in the Table 24. Due to the fundamental nature of some non-destructive testing techniques, the classification is not 100% accurate. The accuracy with which anomalies are classified as being internal vs. external is termed identification accuracy and is quantified by establishing a *Probability of Identification (POI)*.

Within the classification of metal loss, in-line inspection vendors will often report more specific information regarding the nature of the metal loss. That is, the anomalies are further classified based on the inferred dimensions as shown in Figure 29.

deformation: Generally, no section of line pipe is perfectly circular; it will have some out-of-roundness or ovality. This ovality is generally acceptable if it remains within the limits defined by standards, such as API 5L for new construction, otherwise it will cause fit-up problems.

However, if the physical distortion of the pipeline is significant enough, there is a concern regarding unacceptably high localized stresses, safe passage of inspection tools or increased susceptibility to other forms of damage, such as cracking or corrosion. In these cases, (clearly) generally accepted limits have been exceeded and each situation must be addressed on a case-by-case basis. Examples of common types of pipeline deformation that can be a significant concern include dents, wrinkles, and buckling.

Table 24. Typical defect classification categories for an MFL pig run (124)

Classification	Description
Metal Loss	• Usually defined as internal/external. • Also identified as box or cluster (if interaction rules are applied).
Girth welds	• Girth welds will be identified as will associated wall thickness changes.
Mill anomaly / construction defect	• In some cases, location and geometry of an anomaly can be used to determine if the defect is potentially a (non-injurious) construction or mill defect (as opposed to metal loss requiring investigation).
Deformation (dent)	• Deformations (especially in the form of dents) can be identified although sizing of anomalies can become more difficult depending on the technology used.
Casing	• Metal casings – often used at road and railway crossings – can be detected by magnetic techniques.
Sleeve	• Sleeves can usually also be detected depending on the sleeve type and inspection technology used (i.e., magnetic techniques will not detect non-metallic sleeves).
Appurtenances	• Taps, tees, valves bends, flanges, above ground pipe supports, launcher/receiver.
Close metal object	• Magnetic techniques will detect metal objects, not readily identified, in the vicinity of the pipe (e.g., welding rod).

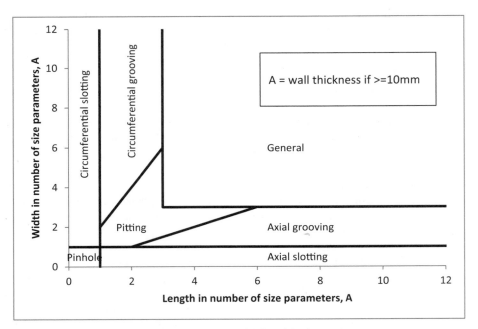

Figure 29. Classification of corrosion anomalies (defects) (16).

deformation tool: A type of smart pig used to assess the nature of the pipeline geometry - usually prior to undertaking an inspection for corrosion or cracking. The term is used interchangeably with *geometry tool* and *caliper tool*. See *pig* for a detailed discussion of this class of tools.

dent: A geometric deformation of the pipe resulting from impact with equipment, rock or similarly large/hard objects. Dents are a severe form of mechanical damage and are a concern from a pipeline integrity point of view as they shorten the fatigue life of the pipeline and, if they are associated with other features, lower the failure pressure. They also restrict the travel of pigs and restrict product flow in the pipeline.

While it has been shown that dents up to 8% deep may not impact burst pressure, most standards use 6% as the acceptable limit. Some shallow dents caused by impact and subsequent re-rounding of pipe can be associated with hardening of the steel. These dents are prone to cracking. When assessing dents, the shape of the dent (smooth) vs. other potentially associated features must be considered -- e.g., the presence of a gouge (localized hardening) or the difficulty in detecting/sizing corrosion that may be present).

Gouges are the greatest concern, where a shallow layer of hardened material may form due to mechanical energy and high cooling rate. When the dent rebounds, this hardened material (often martensitic) cracks. Coating has been removed, so hydrogen forms as a result of cathodic protection, which leads to further damage. Fatigue life is drastically shortened due to the presence of cracks.

Department of Energy (DOE): An arm of the executive branch of the US government established to advance national, economic, and energy security of the United States. Pipelines are of interest to the DOE because they are critical in the transportation of natural gas, petroleum, and refined products in the US.

Department of Transportation (DOT): An arm of the executive branch of the US government established to regulate transportation in the United States. Pipelines, as a form of transportation of natural gas and petroleum, are within its jurisdiction.

The agency within the DOT that regulates pipelines is Pipeline and Hazardous Materials Safety Administration (PHMSA), created in 2005. Its stated goals are to

- Reduce the risk of harm to people due to the transportation of hazardous materials by pipelines and other modes.
- Reduce the risk of harm to the environment due to the transportation of oil and hazardous materials by pipeline and other modes.
- Help maintain and improve the reliability of systems that deliver energy products and other hazardous materials.
- Harmonize and standardize the requirements for pipeline and hazardous materials transportation internationally, to facilitate efficient and safe transportation through ports of entry and through the supply chain.
- Reduce the consequences (harm to people, environment, and economy) after a pipeline or hazmat failure has occurred.

PHMSA, acting through the Office of Pipeline Safety (OPS), administers the department's national regulatory program to assure the safe transportation of natural gas, petroleum, and other hazardous materials by pipeline. OPS develops regulations and other approaches to risk management to assure safety in design, construction, testing, operation, maintenance, and emergency response of pipeline facilities. Since 1986, the entire pipeline safety program has been funded by a user fee assessed on a per-mile basis on each pipeline operator OPS regulates.

depth of cover: Refers to the distance between the top of the pipeline and the soil surface. Regulations specify minimum depths of cover that will vary depending on the nature of the surface activity anticipated as well as ground conditions. For example, a greater depth of cover is required for road crossings as well as in areas of unstable soils. General depth of cover requirements are on the order of ~1 m and vary as per the conditions previously discussed.

derate: A situation where the maximum allowable operating pressure on a line is reduced relative to original design conditions either on a temporary or permanent basis. The most common instance of this is the temporary reduction in pipeline operating pressure to allow sufficient time to mobilize repair/inspection crews where a defect, thought to pose an immediate safety concern (usually based on in-line inspection data), is suspected. A derate may be imposed on a more permanent basis where an operator or regulator deems that pipeline operation at the design pressure is either not economic and/or safe.

design code: See *code*.

design pressure: The pressure a pipeline is designed to contain using accepted engineering practices:

$$P = \frac{2St}{D} \times F \times L \times J \times T$$

> where P = *design pressure*
> St = *yield strength of the pipe material*
> D = *outside diameter of the pipe*
> t = *wall thickness of the pipe*
> F = *design factor*
> L = *location factor (dependent on the class location)*
> J = *longitudinal joint factor (added factor for some types of welds)*
> T = *temperature derating factor (added temperature for high temperature)*

detection: In the context of pipeline integrity, detection is the sensing of an anomaly or imperfection on a pipeline by a non-destructive testing or by an in-line inspection. The ability of a system to detect a given anomaly is measured by two parameters: the probability of detection (POD) and the probability of false call (POFC).

The probability of detection is the probability that an existing anomaly will be found by an inspection. The probability of false call is the probability that an indication identified by an inspection is in fact not a true anomaly on the pipeline.

Following the collection of inspection data, a signal (S) is analyzed to detect an anomaly. A detection threshold (T) is also set such that if signal strength is greater than a predetermined detection threshold (S > T), then the system reports an anomaly. If the signal is less than the detection threshold (S < T), no anomaly is reported. Thus, POD and POFC are related: as the probability of detection increases, the probability of false call also increases (due to increased levels of background noise). See Figure 30. Given the potential consequence of failing to identify a corrosion anomaly, we might expect that in-line inspection vendors will set the detection threshold to yield a high POD and live with a similarly high value of POFC.

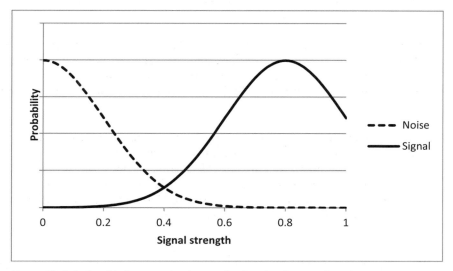

Figure 30. Relationship between background noise signal strength and potential pipe anomalies.

detection threshold: See *detection*.

deterministic: The calculation of any quantity or value of any parameter assuming that it can be known exactly. Deterministic calculations and parameters differ from *probabilistic* calculations and parameters in that probabilistic methods assume that the quantity cannot be precisely known.

dewatering: The removal of free water from a pipeline – usually after conducting a hydrostatic test. The most common technique is to use a series of utility pigs to ensure that the water is absorbed and removed; however, in some circumstances glycol is also used to ensure that water removal is as complete as possible.

diameter: The distance of the longest possible straight line across a cross section of pipe, exclusive of manufacturing tolerances. Pipeline diameter is one of the key considerations during the pipeline design phase due to its direct impact on transportation capacity as well as construction and materials costs.

diameter change: A change in the size of pipe used in a given pipeline segment (often requiring the use of a transition piece). Diameter changes are particularly significant for pigging operations and changes in flow characteristics (particularly where internal corrosion due to erosion is a concern).

dielectric coating: Refers to the range of external pipeline coating formulations that do not conduct electrical current. Thus, this is a generic descriptor for a range of coating types that have this characteristic. See *coating*.

dielectric shield: An electrically nonconductive material, such as a coating or sheeting made from fiberglass, plastic, or other nonconductive material. Shielding is required to improve current distribution and reduce current wastage, usually on the cathode, in cathodically protected locations where the anode and cathode are in very close proximity.

differential aeration cell: An *electrochemical cell*, specifically a *localized cell*, that is created due to differing levels of air (specifically oxygen) in discrete locations on a single structure. An example of this effect is the creation of anodic and cathodic locations on the pipe surface, some distance apart, due to differences in the localized environments (perhaps due to a biofilm or bacterial colony). This term is used interchangeably with the term concentration cell. See *electrochemical cell*.

dig criteria: This term is most often used in the context of assessing anomalies identified through in-line inspection. The criteria are most often a set of defect dimensions (and associated failure pressures) which, if exceeded, dictate that the anomaly must be investigated further given the pipeline is likely at significant risk of failure. One of the most common criteria is that any defect with a depth reported to be $\geq 70\%$ of the pipe wall thickness requires excavation, further investigation, and potential repair. The term is often used interchangeably with *repair criteria* and (erroneously) with *failure criteria*.

diffusion: The spreading of a constituent in a gas, liquid, or solid to a uniform concentration. Specifically, diffusion causes the flow of a substance from areas of high concentration to areas of low concentration. The diffusion rate is governed by several factors:

$$ J = -D \frac{\partial C}{\partial x} $$

where J = diffusion flux rate

   D = coefficient of diffusion

   C = concentration

   $\frac{\partial C}{\partial x}$ = concentration gradient of the substance

A common example of the importance of diffusion in the context of pipeline integrity is the movement of hydrogen, potentially into the pipe wall, contributing to localized embrittlement and other hydrogen damage mechanisms.

diffusion barrier: Any condition, usually a physical obstruction, that represents a barrier to the uniform spreading (i.e., diffusion) of a given substance (gas, liquid, or solid). Modern external pipeline coatings are a common example of a diffusion barrier as they (typically) prevent the uniform distribution of water and oxygen along the pipe surface relative to the ambient environment.

dig criteria: See *repair criteria.*

dig program: A series of excavations that are undertaken on a pipeline – most often as a result of either a condition monitoring program (see *direct assessment*) or as a result of receiving data from an in-line inspection tool. A dig program may be purely investigative as part of direct assessment, be driven by the need to repair potentially injurious anomalies, or be undertaken for calibrating or verifying the results of an in-line inspection.

dig sheet: A reference document, usually a single page, that contains sufficient information for a field crew to locate, excavate, and identify an anomaly identified through in-line inspection. As such, the document will typically include the items cited in Table 25.

direct assessment: A methodology of pipeline integrity maintenance and verification that uses various indirect measurements and surveys to identify locations to excavate and directly examine the pipeline. Table 26 shows several Recommended Practices and Standards published by *NACE International*; each Standard and RP addresses a specific threat. Each of the direct assessment standards follows a four-step process, as shown in Figure 31, with a feedback loop after every step.

Table 25. Typical components of a dig sheet (124)

Dig sheet items	Description
Upstream/downstream control point(s)	Readily identifiable points, upstream and downstream of the anomaly location, providing a common reference point for establishing and comparing the pipeline surveyed *chainage* vs. the chainage reported from the in-line inspection. Where there are significant discrepancies between the pipeline survey chainage and that reported by the in-line inspection tool, multiple control points may be used to resolve discrepancies and locate the anomaly with confidence.
In-line inspection chainage	The chainage of the anomaly to be investigated as identified by the in-line inspection tool.
Survey chainage	The chainage of the anomaly to be investigated based on pipeline survey chainage, i.e., the in-line inspection tool chainage that has been adjusted for any error through the process of matching in-line inspection tool and survey chainage for commonly identifiable control points.
O'clock position	The circumferential orientation of the anomaly on the pipeline when facing the downstream direction on the pipeline.
Upstream girth weld distance	The distance of the closest upstream girth weld.
Downstream girth weld distance	The distance of the closest downstream girth weld.

Table 25, continued

Dig sheet items	Description
Coating type	Indication of the expected coating type to increase confidence in locating the anomaly as well as ensuring that field crews are adequately prepared in case special handling procedures are needed (e.g., some coatings contain asbestos and must be handled accordingly).
Pipe diameter and wall thickness	Identification of the expected pipe diameter and wall thickness to ensure that the anomaly has been correctly identified.
Identification of other anomalies in close proximity	Indication of any other anomalies expected in the vicinity to ensure that the anomaly has been correctly identified.
Other significant pipe features	Where possible, identification of any other features on the pipeline that will assist with the location of the anomaly (e.g., branch connections, meter locations and so forth)

Table 26. NACE Standards associated with direct assessment (184)

Standard No	Title
SP0502-2008	Pipeline External Corrosion Direct Assessment Methodology (ECDA); formerly RP 502
SP0206-2006	Internal Corrosion Direct Assessment Methodology for Pipelines Carrying Normally Dry Natural Gas (DG-ICDA)
SP0204-2008	Stress Corrosion Cracking (SCC) Direct Assessment Methodology (SCCDA); formerly RP204
SP0208-2008	Internal Corrosion Direct Assessment Methodology for Liquid Petroleum Pipelines (LP-ICDA)

Figure 31. Four-step direct assessment process.

direct current (DC): A specific type of *electrical current* where the flow of electrons is consistent in one direction This is in contrast to alternating current where the flow of electrons reverses on a regular cycle consistently in one direction (such as from batteries). From the point of view of the power grid, AC power experiences fewer transmission losses and necessary voltage changes are much cheaper to do.

From the point of view of non-destructive testing, AC current has very different characteristics (vs., DC) when traveling through metal (i.e., the current travels largely on the surface of the material vs., the entire cross section of the material) making it more suitable for certain applications such as detecting surface breaking cracks. AC also generates a correspondingly varying magnetic field – a phenomenon that has been explored and utilized in a number of non-destructive technologies.

direct current voltage gradient survey (DCVG): A technique that measures the rate of change of electrical potential of the ground. The survey is very similar to an *alternating cur-*

rent voltage gradient (ACVG) except that it uses a direct current power source instead of an alternating current power source.

DCVG surveys are used to detect coating *holidays* and *disbonded coating* on pipelines with cathodic protection. The survey uses a pulsed DC power source to put an electrical current into the ground. In theory, a coating holiday creates an electrical voltage gradient in the surrounding soil because it acts as a source or sink for the current. The electrical potential will increase (or decrease) in all directions around the holiday.

direct examination: See *direct assessment.*

disbonded coating: A type of coating failure where there is a loss of adhesion, or a detachment, between the coating and the surface of a pipeline. The disbondment of a coating can allow water and other corrosive environments to come into contact with the pipe surface and cause corrosion – an effect that is exacerbated if the disbondment is extensive and the nature of the failure is such that the surface of the pipe is shielded from *cathodic protection.*

disbondment: See *disbonded coating.*

discrete mitigation: The process of executing on a set of activities – along discrete lengths of the pipeline – to reduce the risk exposure to a particular pipeline integrity threat. An example of discrete mitigation would be the excavation and remediation of an in-line inspection tool identified anomaly on the pipeline vs. cathodic protection, which is applied to the entire length of the pipeline. See also *mitigation.*

distribution line: A gas pipeline used to deliver gas directly to customers. Distribution lines can be considered low pressure or high pressure. Low-pressure distribution lines operate at service pressure (i.e., the same pressure that end-users such as homeowners use to run appliances, etc.) and therefore do not need a regulator to reduce pressure at the end of the line. High-pressure distribution lines operate at a pressure higher than delivery pressure and require a regulator just prior to custody transfer.

Other characteristics of distribution lines are that they are often constructed within road allowances in urban (Class 4) locations. Also, the product is odorized to support early detection of pipeline leaks, and the pipeline tends to have a relatively high number of customer off-takes and mix of pipeline diameters. The pipe material varies much more than gathering and transmission pipe to include steels, plastics, and composites. The primary failure mode is leakage because the pipelines operate at significantly lower stress levels (<30% *smys)*.

As such, the pipeline integrity programs associated with distribution systems can be quite different than those that would be considered appropriate for large diameter transmission or gathering systems. The key differences such as product quality, pipeline material, diameter, and consequence of failure can drive vastly different decisions in the area of inhibitors, inspection programs and repair vs. replace decisions. See also *pipeline.*

documentation: The formal record keeping, either hardcopy, electronic or otherwise, that can be used to establish an audit trail to establish the course of events and actions that may have occurred. In the context of pipeline integrity, these records include, but are not limited to, items such as drawings, maps, inspection records, failure reports, as well as the results of ongoing monitoring activities.

DOE: See *Department of Energy.*

DOT: See *Department of Transportation.*

double block and bleed: Refers to the capability of a valve, under pressure, to achieve a seal on both the upstream and downstream seats with the body cavity evacuated to atmospheric pressure. This requires a drain (oil lines) or a vent (gas lines) to discharge product from the body cavity. Fitting a gas detector to the port provides assurance of the integrity of the upstream seal. This configuration is often required to isolate high-pressure sections of a system to facilitate safe maintenance, etc. Where double block and bleed cannot be accomplished with a single valve, a section of pipe between two valves maybe used to obtain a similar level of safety.

double submerged arc weld (DSAW): A variant of the *submerged arc weld.* Refers to the longseam weld in pipe manufactured by the use of two submerged arc welds to form the longitudinal seam. See *weld.*

downstream: Indicative of direction along the longitudinal axis of a pipeline relative to a given reference point that may be located anywhere along the line; specifically, the downstream direction is away from the point from which the product flows.

drag-reduction agent: An additive that results in the reduction of frictional losses between the pipe wall and product – ultimately resulting in an increase in transportation capacity of the system. There are two main types: high polymers and surfactants – the choice of either being a function of the specific situation. Drag reduction agents work best where the majority of frictional losses result from turbulent flow in (mostly) straight pipe. The first commercial demonstration of the success of drag-reducing agents was on the Alaska pipeline in 1979 where pipeline capacity was increased by ~25%.

drainage: 1. In the context cathodic protection, drainage is the conduction of electric current from an underground metallic structure by means of a metallic conductor. Cathodic protection systems essentially rely on *forced drainage*; that is the application of an electromotive force, such as the connection of a sacrificial anode, to ensure that there is an excess "supply of electrons" to the pipeline to prevent surface corrosion. **2.** See also *soil drainage.*

drainage bonds: An electrical circuit established, as part of a cathodic protection system, to conduct unwanted electrical current away from the pipeline being protected. Specifically, drainage bonds are most often used to address issues associated with stray currents induced by structures such as power lines and railways in the vicinity of the pipeline being protected. See also *forced drainage*.

drip leg: A low point bypass and pressure vessel installed on gas pipelines at regular intervals to capture any liquids (either out of specification product, valve grease etc.,) and debris in a manner that minimizes disruption of the pipeline flow. See Figure 32.

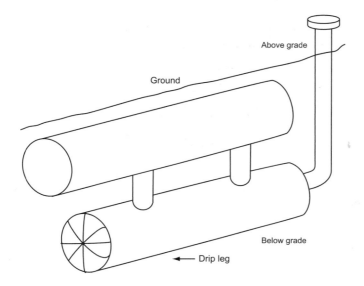

Figure 32. Drip leg used to collect liquids in a gas line.

drop weight tear test (DWTT): An impact test commonly used to assess the resistance of pipeline steel to fracture propagation (i.e., fracture toughness). The DWTT uses a full thickness test specimen with a 5-mm-deep pressed notch in the center of the span. Once mounted in the test apparatus, the specimen is fractured in a single impact by a falling weight (on the side opposite to the notch). See Figure 33. The key parameters measured through this test include the absorbed energy (based on the impact test) and the percentage of shear area on the fracture face. Experiments have shown a correlation between DWTT, transition temperature, and fracture speed, as well as between the shear area of a DWTT specimen and the fracture speed. Thus, the DWTT can be used to establish the pipe material's resistance to fracture propagation. This term is used interchangeably with the term pressed-notch DWTT specimen.

Standard Test Method for Drop-Weight Tear Tests of Ferritic Steels

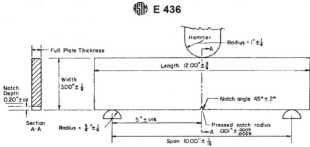

E 436

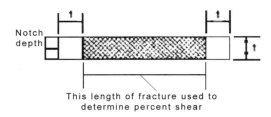

Figure 33. Schematic of drop weight tear test. (198)

driving potential: The difference in electrical potential between an anode and a steel structure that is being protected.

DSAW: See *double submerged arc weld.*

dual-diameter pig: A *pig* designed to operate in two different pipe diameters. The development of dual- (and multi-) diameter tools has been in response to many (older/subsea) pipelines that have sections with different diameters. Dual-diameter operation is not a standard configuration for most pigs, and vendors must be included in assessing if (and how) a dual-diameter line is to be pigged. Table 27 shows key factors in pigging dual-diameter lines.

Table 27. Key considerations in designing a dual-diameter pig (124)

Parameter	Characteristics
Nature of transition	• Abrupt transitions present a greater concern because the tool could be damaged if it "impacts" the transition fittings at high speed.
Flow conditions	• Low pressure/flow conditions can increase the potential for tool damage at the diameter change as the pig may initially stop and then experiences an uncontrolled velocity excursion as pressure builds up behind the tool causing it to "break free."
Pigging technology	• More sophisticated (smart) pigs tend to have less flexibility in negotiating diameter changes due to the nature of the (relatively fragile) hardware. • Cleaning and other utility pigs are typically easier to adapt and less prone to damage.

ductile fracture: In a structural engineering context, fracture characterized by tearing of metal accompanied by appreciable gross plastic deformation and expenditure of considerable energy. Ductile fractures require significantly more energy than brittle fractures and are a more "desirable" mode of failure because the likelihood of a running fracture is significantly lower. Specifically, ductile fractures are more likely to run below the speed of the (methane) decompression wave. It is still possible to experience a running fracture because the gas expanding in the bulging pipe immediately behind the crack tip provides significant amounts of energy for propagation. See also *fracture control*.

ductile tearing: See *ductile fracture*.

ductility: The ability of a material to deform plastically without fracturing; it is measured by elongation or reduction of area in a tensile test, or by other means, and is a critical parameter when choosing a pipeline material. See *stress-strain curve*.

dummy pig: a type of utility pig usually run to assure the safe passage of an intelligent pig, to act as sealing pig for *hydrotesting*, or for cleaning to remove accumulated solids. Often used interchangeably with *utility pig* (e.g., non-intelligent tool). See *pig*.

duplex steel: A class of steels that is often specified for use in situations where higher corrosion resistance is required (such as sour service). These steels usually have specific microstructure and alloying requirements as determined by a *pitting resistance equivalent number (PREN)*. Further guidance can be found in NACE MR0175. See *pitting resistance equivalent number* and *super duplex*.

DWTT: See *drop weight tear test*.

dynamic testing: Dynamic testing is a form of *pressure testing* defined and discussed in API 1110. Specifically, a form of pressure testing where the pipeline (liquid) product is used to confirm the operating pressure of a line by increasing the pressure to 110% of the MAOP of the line for a minimum of 2 hours.

E

eccentric casing: Describes a situation where the pipeline is no longer centrally located with reference to its casing. This can occur where there has been ground movement or any other factor that could result in the loss or damage to the spacer(s). The most significant concern from an integrity point of view is the potential for the casing to make physical contact with the pipeline wall and result in a cathodic protection short.

ECDA: See *external corrosion direct assessment.*

ECDA region: See *external corrosion direct assessment.*

eddy current inspection: A non-destructive testing method to find cracks and other anomalies located on or close to the surface being inspected. The method is restricted to the inspection of electromagnetically conductive materials and is limited in terms of component depth.

An eddy current inspection requires a varying magnetic field, for example from a coil. Typical frequencies range from below 1 kHz to 10 MHz. The depth of inspection decreases with frequency because of the skin effect. That is the conductors and especially ferromagnetic material are such that the effect of the varying magnetic field is confined to a (relatively) thin layer (skin) along the surface of the material. The alternating magnetic field induces a current in the conducting material and the magnetic field, resulting from the induced current, is measured by sensors at the surface. The induced current is unable to flow through discontinuities in the material and diverts the flow of current – thereby changing the magnetic field measured by the sensors. Changes in the magnetic field are detected by the sensors and interpreted as discontinuities or flaws in the material.

elastic deformation: The non-permanent deformation that a material undergoes due to strain. Following the release of the load, the material returns to its original dimensions without any permanent deformation. If the stresses exceed the elastic limits, then the material will undergo *plastic deformation* or, possibly, *brittle fracture.* See *stress-strain curve.*

elastic modulus: See *Young's modulus.*

elastic wave: Mechanical waves that travel through solids. They include several specific types of waves, called modes. These wave modes include body waves, such as longitudinal and transverse wave, and surface waves, such as *Love* and *Rayleigh waves*. The term elastic wave usually implies that the wave is traveling through a solid medium rather than a gas or liquid; however, the longitudinal wave mode will pass through fluids. See *ultrasonic testing*.

elastic wave tool: A type of in-line inspection tool for detecting cracking in gas pipelines. The technology uses specialized liquid-filled wheels to transmit the ultrasonic impulses into the pipe wall. The use of the wheels eliminates the need for a liquid coupling to transmit the ultrasonic signal into the steel.

electrical current: Electrical current is the amount of electrical charge flowing between two points per unit time. Current is measure as a ratio:

$$i = q / t$$

where i = current
    q = net flow of electrical charge
    t = time

The standard SI unit for current is the ampere (A); 1 ampere is 1 coulomb per second. Small currents are often measured in milliamps (mA) (10^{-3} A) or microamps (μA) (10^{-6}A).

electrical field signature method: This is a method of assessing corrosion damage and monitoring corrosion. Unlike many other corrosion monitoring methods, it is a permanent installation at a facility where it measures corrosion damage over a large area of several meters; most other methods use small (portable) sensors to monitor corrosion at a specific point.

The electrical field signature method uses Ohm's Law to calculate the thickness of a metal structure. Following that, an electrical current is applied to the structure and the electrical potential of the surface of the metal is monitored by an array of sensors. The electrical potential at each point is dependent on the geometry and condition of the structure at the time of the measurement. The potentials at the time of installation are recorded and saved as the signature. Any change in that signature indicates some change in the condition of the structure. The potential at the sensor locations is used to calculate the current density and resistance between the points, which can then be related to the thickness of the metal. The method is often used to monitor internal corrosion, where the sensory array can be placed on the external surface of a pipeline.

electrical isolation: This describes a metallic structure that is electrically isolated from the surrounding environment and/or other structures in the vicinity. The concept of a structure being electrically isolated is significant in the context of cathodic protection systems, where all metallic structures that are electrically connected have the potential to impact the cathodic protection circuit and corresponding protection levels (the impact may be either positive or negative depending on the specific situation).

electrical potential: The amount of energy per unit charge. Absolute potential energy is not meaningful, therefore the electrical potential is always measured between two points. The potential energy is the amount of work required to move a unit charge between the two points.

electrical resistance corrosion monitoring (ER): This is a method of measuring wall thinning of a metal based on its electrical resistance. Electrical resistance is inversely proportional to the cross-sectional area of the material; thus, as a metal corrodes, the cross-sectional area decreases and the electrical resistance increases. The resistance is compared to an uncorroded reference sample, which compensates for temperature variation. The ER probes are permanently attached to the external surface of the pipeline and can be monitored manually on a periodic basis or continuously from a centralized location. Due to the nature of ER monitoring, the technique does not distinguish between localized and general corrosion, and it is susceptible to error from conductive corrosion deposits.

electric resistance weld (ERW): A weld formed by passing an electrical current through the metal to be welded. The heat is generated from the electrical resistance at the two metal edges. The edges are force together to create a weld. Older (low frequency) ERW pipe was of poorer quality and its use was primarily for low pressure water lines. New (high frequency) ERW pipe has vastly improved in quality and is now used for high pressure gas pipeline applications. See also *welding*.

electrical shielding: The disruption of electrical current flow due to the presence of a (electrically) non-conductive material. See *shielding*.

electrochemical cell: A system containing the four key components for creating an electrical circuit: an anode, a cathode, an electrically conductive medium (i.e., electrolyte), and a conductive path between the anode and the cathode. The *anode* is the electrode of an electrochemical cell at which oxidation occurs (i.e., metal loss). The *cathode* is the electrode at which reduction occurs. Electrons flow through the external circuit from the anode to the cathode. Corrosion usually occurs and metal ions enter the solution at the anode. It is important to note that the cathode and anode can exist as discrete locations on the same structure, or involve otherwise independent structures. See Figure 34.

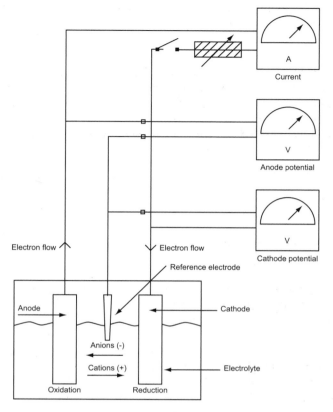

Figure 34. Anode and cathode in a galvanic cell (18).

electrode: A conductor (i.e., an *anode* or *cathode*) used to establish contact with an electrolyte (e.g., soil) to complete an electrical circuit. See *electrochemical cell* and cathodic protection.

electrode potential: The voltage difference of an electrode in an electrolyte as measured against a reference electrode. The measurement does not include any losses due to the resistance in the remainder of the electrical circuit, therefore representing the reversible work to move a unit of charge from the electrode surface through the electrolyte to the reference electrode.

electrolyte: A (usually aqueous) solution that contains ions that facilitate the flow of an electrical current.

electromagnetic acoustic transducer (EMAT): A type of ultrasonic transducer that does not require physical coupling to the material to be inspected. Specifically, a permanent magnetic is used to establish an underlying magnetic field in the pipe. Added to this, a coil, oriented 90° to the magnetic field, carries an alternating current that induces eddy currents in the metal. The interaction of the eddy currents and the underlying magnetic field generate compression waves in the pipe wall (see *ultrasonic testing*). This technology is of particular interest for detecting cracks (or crack like defects) in gas pipelines where good coupling for the transmission of ultrasonic waves is expensive to establish.

electromotive force series (EMF): This is a list of elements arranged according to their standard electrode potentials as compared to hydrogen. The potential is positive for elements whose potentials are cathodic to hydrogen and negative for those anodic to hydrogen. The EMF series can be used to predict whether one metal will corrode if it is in contact with another (dissimilar) metal, based on whether the two metals are cathodic or anodic with respect to another.

eluviation horizon: See *soil horizon*.

EMAT: See *electromagnetic acoustic transducer*.

EMAT tool: This is a type in-line inspection tool that uses EMAT technology to inspect (primarily) gas pipelines for the purposes of detecting cracking and crack-like defects. See also *electromagnetic acoustic transducer*.

embrittlement: Refers to a material undergoing a loss of ductility due to a chemical or physical change. A common example in pipelines is the migration of hydrogen to localized areas of stress causing localized loss of ductility (*hydrogen embrittlement*).

emergency response plan: A document that establishes a basis for the coordinated and timely response to an incident to minimize the impact to the public, the environment, and company operations. While the existence of an emergency response plan is a regulatory requirement, it is also an important way of mitigating potential financial risk for the pipeline owner.

emergency shutdown: A halting of pipeline operations due as a result of an adverse event or in response to the imminent threat of an adverse event occurring. Depending on specific context (i.e., product, facility, nature of alarm, etc.,), pipeline product may also be evacuated from the facility.

emergency shutdown valve: An automated (see *actuator*) valve that is triggered to close when a dangerous situation is detected in order to minimize harm to people, property or the environment. The most notable examples include the valves on a gas pipeline system that cycle closed when a significant rapid drop in pressure (i.e., potentially indicative of a rupture) is detected.

EMF series: See *electromotive force series*.

enamel: Refers to the range of external pipeline coating formulations, usually asphalt or coal tar based, that upon drying/curing form a hard glassy surface. See *coating*.

encroachment: A situation where location of third party activities, facilities, or structure infringe upon the pipeline right-of-way. It is of particular concern in terms of *mechanical damage*.

endurance limit: The maximum alternating stress level below which a material can withstand an infinitely large number of fatigue cycles without failing. Figure 35 illustrates the endurance limit on an S-N (stress versus number of cycles) diagram. As the alternating stress level decreases, the number of cycles to failure increases. Some materials can withstand an unlimited number of cycles if the stress is below some threshold. That threshold is the endurance limit.

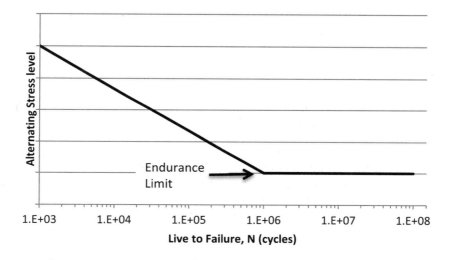

Figure 35. Endurance limit on an S-N diagram.

engineering assessment: A situation-specific assessment. In the context of analyzing a defect on the pipeline, the analysis is typically undertaken as part of determining if the defect is injurious and/or determining the optimal repair procedure.

environmentally assisted cracking (EAC): Environmentally assisted cracking is a general term referring to mechanisms that cause (localized) brittle fracture of normally ductile materials. Some of the more common types of EAC includes corrosion fatigue, hydrogen embrittlement, hydrogen-induced cracking, hydrogen stress cracking, stress corrosion cracking, and sulfide stress cracking.

Environmental Protection Agency (US): Created in 1970 in response to the growing public demand for cleaner water, air, and land, the EPA was established to make a coordinated attack on the pollutants that harm human health and degrade the environment. The EPA was assigned the task of repairing the damage already done to the natural environment and to establish new criteria to make a cleaner environment a reality.

EPA (US): See *Environmental Protection Agency.*

epoxy: Refers to the range of external pipeline coating formulations based on resin(s) formed by the reaction of aliphatic or aromatic polyols. Other uses include under repair sleeves or as grouting. See *coating.*

epoxy coal tar: Refers to the range of external pipeline coating formulations based on a combination of resin(s) formed by the reaction of aliphatic or aromatic polyols and coal tar. See *coating.*

EPRG: See *Eurpoean Pipeline Research Group.*

ERF: See *estimated repair factor.*

erosion: 1. The loss of material due to the impact of particles on the internal pipe surface. There are three key mechanisms for erosion, in all cases higher fluid velocities exacerbate the problem – particularly at locations that disturb fluid flow (e.g., bends, high weld beads, diameter reductions). Obviously the larger the number of particles, as well as larger particles, will also increase erosion. Table 28 shows the three key mechanisms causing erosion in a pipeline. **2.** Erosion is the breakdown and subsequent movement of rock, soil, backfill, or other natural material due to the action of gravity, ice, wind, or water on the surface of the earth. Erosion can be a particular concern for pipeline integrity because it can result in several undesirable situations, such as slope instability (i.e., potential for significant ground movement), pipe loss of cover or floating pipe.

erosion-corrosion: The combined effect of an electrochemical reaction of a metal or alloy and the effects of mechanical wear or abrasion. Erosion corrosion is often found in piping at elbows, bends, and joints where the direct of flow is changed.

ERW: See *electric resistance welding.*

Table 28. Three key mechanisms in internal pipe wall erosion (10)

Mechanism	Description
Impact of liquid drops or jets	• Can result in the creation of shallow craters or longitudinal features in ductile pipe
Impact of particles	• Typically a particle size > 5 µm causes erosion concerns • The greater the number of particles, the greater the erosion concern
Cavitation	• The mechanism involves growth and collapse of bubbles due to localized pressure fluctuations in a liquid; the collapse is accompanied by a rapid flow of liquid and stress fluctuations at the pipe surface causing erosion

estimated repair factor (ERF): This is a ratio of the B31G failure pressure of a pipeline anomaly relative to the maximum allowable operating pressure of the pipeline. Thus, an ERF <1 would indicate an anomaly is acceptable by B31G assessment method; conversely, an ERF > 1 indicates an anomaly that requires repair. Thus:

$$ERF = \frac{MAOP}{P'}$$

where $MAOP$ = maximum allowable operating pressure

P' = safe operating pressure (1.1 x failure pressure)

European Pipeline Research Group (EPRG): Established in 1972 by a group of pipe manufacturers and transmission companies. The main activities of this group have been focused on the managing the threat of pipeline leaks and ruptures as well as the development of higher strength steels and performance standards for new pipelines. Over the course of 40 years, EPRG has become grown to include 17 member organizations responsible for pipe manufacturing capacity of ~ 2 million tons / year and the operation of over 100,000 km of high-pressure gas transmission pipeline. (193)

excess cap height: See *excess convexity*.

excess concavity: This is a weld flaw where the cap or root of the weld has a concave profile. See Figure 36.

A depression in the top of the weld, or cover pass, indicating a thinner than normal section thickness.

excess convexity: This is a weld flaw where the cap or root of the weld has an excessively convex profile. Usually, a cap that exceeds a height of 3 mm above the parent metal is considered to be excessive. See Figure 37.

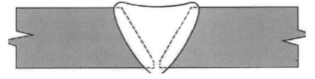

Figure 36. Excess concavity in a weld. (197)

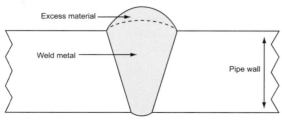

Excess Convexity

Figure 37. Excess convexity in a weld.

excess penetration: A welding defect that occurs when too much heat is present in the weld area and the weld metal penetrates too far into the base metal. The excessive heat can be the result of either poor technique (lack of speed) or incorrect (i.e., excess) volts and amps input. In the more extreme cases, metal may sag through the weld leaving a gap in the weld area. See Figure 38.

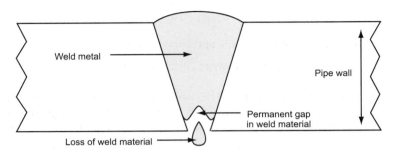

Figure 38. Excess penetration in a weld.

exchange current: The rate at which negative charge is leaving the surface when an anode reaches dynamic equilibrium in an electrolyte. The exchange current, usually expressed as the current per unit area, is directly proportional to the weight loss of due to corrosion. The exchange current density can be a key factor in selecting anode materials because high current exchange density ultimately results in a more efficient cathodic protection system.

exponential distribution: A special case of gamma distribution. Mathematically, the probability density function of the gamma distribution, $ex(x)$, is defined as

$$ex(x) = \begin{cases} \lambda e^{-\lambda x}, & 0 < x < \infty \\ 0, & x \leq 0 \end{cases}$$

where λ = parameter of the distribution

The CDF, $Ex(x)$, is given by

$$Ex(x) = 1 - e^{-\lambda x}$$

Table 29 shows selected properties of the exponential distribution. Figure 39 shows the shape of the exponential distribution for various values of λ. The exponential distribution has been used to model the distribution of corrosion rate on pipelines.

Table 29. Exponential distribution properties

Exponential distribution properties	
Parameters	λ
Domain	$0 \leq x \leq \infty$
Mean	$1/\lambda$
Variance	$1/\lambda^2$
Mode	0

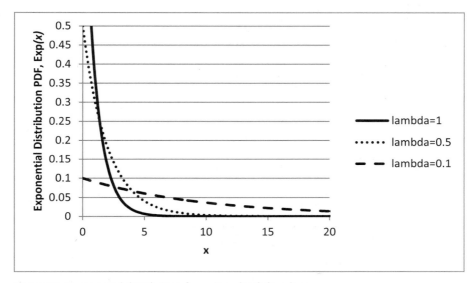

Figure 39. Exponential distribution for various lambda values.

external corrosion: Corrosion encountered on the outer surface of the pipe. See *corrosion*.

external corrosion direct assessment (ECDA): This is a supplemental method for assessing pipelines (i.e., identifying susceptible areas where corrosion activity has occurred, is occurring, or may occur) in high consequence areas, where the pipeline operates at low stress or to assess mechanical damage. ECDA can also be used as the primary method for assessing pipelines where hydrostatic testing or in-line inspection is not feasible. ECDA is a structured four-step process articulated through NACE Standard Recommended Practice on Pipeline External Corrosion Direct Assessment Methodology (RP0502-2002) and enacted into law in the US by 49 CFR Part 192 Rule (Pipeline Integrity Management in High Consequence Areas (Gas Transmission Pipelines)). Figure 40 shows a simplified flow chart of the ECDA process.

extruded polyethylene jacket material: See *coating* (specifically, Table 16).

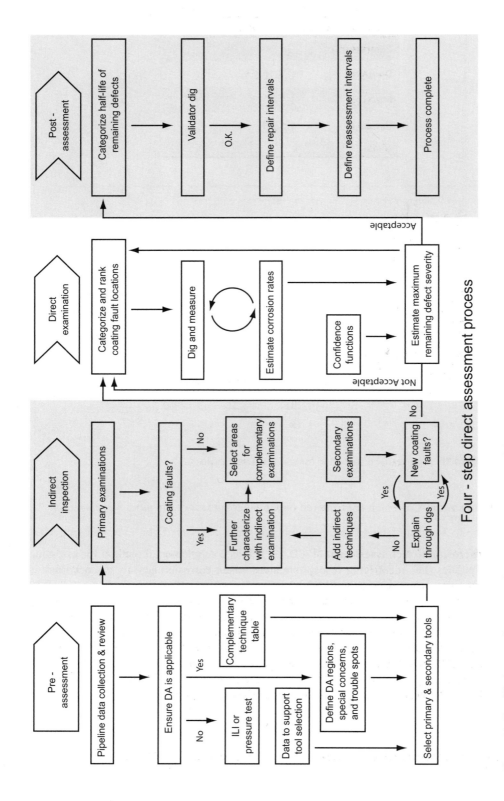

Figure 40. Simplified external corrosion direct assessment process (3).

F

facultative anaerobes: Organisms that can survive in the presence and the absence of oxygen in the ambient environment. The most common examples of these include yeasts and sulfate-reducing bacteria.

failure: The point at which a structure is no longer capable of serving its intended purpose. In the context of pipeline integrity, the most common manifestation is the uncontrolled loss of product that results when either a leak or rupture occur.

failure consequence: See *consequence*.

failure criteria: This term is most often used in the context of assessing defects found on the pipeline. The criteria are most often a set of defect dimensions (and associated failure pressures) which, if exceeded, dictate that the defect must be repaired given the pipeline is likely at significant risk of failure. Minimum standards for failure criteria are typically defined within the applicable codes and standards and may be modified (to be more conservative) based on corporate policies. One of the most common criteria is that any defect with a depth $\geq 70\%$ requires some form of repair. The term is often used interchangeably with *dig criteria* and *repair criteria* – two terms that typically refer to potential defects (i.e., anomalies) as identified and reported through in-line inspection (vs., in-the-ditch measurement).

failure mechanism: The set of processes or conditions that compromise the ability of a structure to serve its intended purpose. For example, common failure mechanisms for a pipeline include stress corrosion cracking, corrosion, mechanical damage, etc.

failure mode: The failure mode refers to the way in which a structure fails and how it impacts the operations of the system. In the context of pipeline integrity, three failure modes are often considered: they are leak large leak and rupture. Each of these failure modes result in the uncontrolled loss of product to the environment. It should be noted that gross plastic deformation (i.e., such as buckling), without product release, is also considered a mode of failure since the ability of the pipeline to withstand future stresses and stresses (within design tolerances) has been compromised.

failure pressure: 1. The actual pressure of the pipeline when a failure occurs. This value can be an important part of the failure analysis process providing insight into the exact nature of the failure (and associated failure mechanism(s). **2.** The failure pressure is a calculated value at which a defect is expected to fail based on the available information on the size and properties of the pipe and the defect sizing information from either field measurement, corrosion or crack detection tools.

false call: The reporting of an anomaly, by an in-line inspection tool, that does not actually exist on a pipeline.

fatigue: Refers to failure that is largely induced by continued cyclic stresses beyond the fatigue limit (but less than the maximum tensile strength) of the material.

fatigue cracking: Cracking induced by *fatigue*. See *fatigue*.

fatigue strength: The greatest cycle of stresses that can be tolerated under dynamic loading in a particular context. The strength is specified in terms of the number cycles for specific components. The fatigue strength varies significantly based on component size/geometry, the nature of the loading (i.e., if strain is largely limited to the elastic region or not), surface finish/treatment, temperature, and other environmental factors. This term is often used interchangeably, although not strictly correctly, with the term *endurance limit*.

fault current: See *stray current*.

FBE coating (FBE): Refers to the range of external pipeline coating formulations based on resin(s) formed by the reaction of aliphatic or aromatic polyols. Specifically, the resin, in powder form, is sprayed on the hot pipe surface to form the protective layer. It is one of the most common external pipeline coatings in use today. See *coating*.

FCAW: Refers to flux cored arc welding. *See weld.*

FEA: Finite element analysis. See *finite element modeling*.

feature: Any part or anomalous indication detected by an in-line inspection tool during the performance of an inspection run. Features may be indications of corrosion, girth or seam welds, pipeline valves and fittings, nearby metal objects, or other items to which the technology used is sensitive.

FEM: See *finite element modeling*.

fibreglass sleeve: A hollow fibreglass cylinder encircling the pipe for the purpose of repairing a known anomaly or flaw in the pipeline.

field bend: A bend in the line pipe that is done at a field location – typically without access to heat treatment equipment – limiting the *radius of curvature* that can be introduced (without exceeding acceptable limits to cold work and wall thinning of the pipe material). Typically field bends cannot be fabricated to be any tighter than 5D.

field signature method (FSM): See *electrical field signature method.*

fill rate: The volume of water flowing, per unit of time, during a hydrostatic test to fill the section of pipe to be tested.

fill(ers): The filler passes are the layers of weld material deposited between the second layer (hot pass) and the final layer (cap) of weld material. See *weld passes.*

final report: The term most often used in the context of pipeline integrity to refer to the definitive report on pipeline anomalies provided by the in-line inspection vendor. This is in contrast to any interim or *preliminary in-line inspection report* a vendor may provide. The contents of the final (and preliminary) report should be specified in the *reporting requirements.* The specifics of reporting times and size/nature of defects identified vary significantly based on the specific contractual agreement, *reporting requirements,* the in-line inspection technology used, the density of anomalies in the pipeline as well as its length and diameter. However, typical reporting times for final reports are on the order of 60 days.

finite element modeling: A sophisticated mathematical technique for solving the underlying differential equation by an approximation. Continuous surfaces are discretized by dividing the area (i.e., domain) to be analyzed into discrete sections known as elements – creating a "mesh" structure. Applying conditions of continuity or equilibrium to the model, contact points between the elements result in simultaneous equations which can be solved numerically. This approach can be applied to a range of problems including modeling of electromagnetic fields, stress fields, and fluid flow. One of the key advantages of this approach is the ability to establish very fine meshes in the areas of a component that are of particular interest while reducing the need for intensive mathematical calculations for other areas of the component deemed less critical. Thus, the approach can facilitate a very detailed understanding of localized phenomenon that may be involved with situations such as the impact of a crack or dent on the stress field within the pipe wall. The term finite element modeling is often used interchangeably with the term finite element analysis. Significant resources for this method of mathematical modeling can be found through the National Agency for Finite Element Methods and Standards (NAFEMS) – a not-for-profit UK based body (http://www.nafems.org).

fish eyes: areas on a steel fracture surface having a characteristic white, crystalline appearance.

fitness for service: A methodology for determining that a pipeline can be operated safely based on detailed, context specific, engineering analysis. A fitness for service analysis is a staged approach, beginning with a simple, but conservative, assessment. Depending on the results of the assessment, increasingly complex assessments are conducted until a final determination is made. The term is most often used in reference to situations where a pipeline failure has occurred or an unusual pipeline anomaly, not suitable for assessment through more generally accepted practices, has been detected.

flash welding: A term, often used interchangeably with the term butt welding, that refers to a welding process where the surfaces are forced together while a current passes through them (a form of electric resistance welding). The result is rapid heating of the weld area and a discharge of some of the surface material (i.e., "flash") as the components are brought together. This process ensures that the loss of surface material through the welding process clears undesirable contaminants and provides a sound weld. The extraneous material, or "weld upset," is usually removed by shearing or grinding. The process was used in Russia and other FSU countries for girth welding.

floating pipe: Pipe that has moved to the surface (and is likely to be exposed) due to a buoyancy issue. The issue may result from either a change in site conditions or if the buoyancy measures have been compromised in some way (i.e., weights, originally placed on the pipeline, shift or slide from the desired position).

flow-improvement agent: See *drag-reduction agent.*

flow regime: Describes the character of the flow of a gas and/or liquid. Flow regime is a function of the rate of flow, viscosity, density of the fluid(s), mixing characteristics, and the diameter of the pipe. It is usually assessed by the value of the Reynolds Number, Re. A summary of common flow regimes is shown in Table 30. Flow regime is an important consideration for several reasons. It is an important factor in determining where fluids such as water might collect at the bottom of the pipe and cause internal corrosion. Also, flow regime affects the efficacy of various methods of applying internal corrosion inhibitor to a pipeline. See also *multiphase flow*

Table 30. Various single and multiphase flow regimes in pipelines (10)

Flow regime	Description
Laminar	A single-phase flow regime that is regular and smooth – without turbulence. See *turbulent flow*.
Partially turbulent	A single-phase flow regime where the flow is laminar along the internal surface of the pipe but turbulent in the center. See *partially turbulent flow*.
Fully turbulent	A single-phase flow regime where the flow is turbulent across the entire cross-section of the pipe. See *turbulent flow*.
Stratified flow	A multiphase flow regime where the heavier fluids flow in a layer at the bottom of the pipe, and lighter fluids flow in a layer at the top. Liquids will collect in low areas of the pipeline. See *stratified flow*.
Wavy stratified flow	A multiphase flow regime where gas above liquid phase causes waves in the liquid to push it along. See *wavy stratified flow*.
Slug flow	A multiphase flow regime where the waves of the wavy-stratified flow are sufficiently high to reach the top of the pipe creating slugs of liquid. See *slug flow*.
Bubble flow	A high-velocity multiphase flow regime where the gas phase becomes entrained in the liquid. See *bubble flow*.
Annular mist	A high-velocity multiphase flow regime where the liquid phase becomes entrained in the gas. See *annular mist*.

flow stress: The stress required to maintain the continued plastic deformation of a component (i.e., after it has yielded) but before it reaches its ultimate tensile strength. It is a key input parameter used in corrosion assessment equations assuming ductile fracture, such as *B31G* and *RStreng*.

For the B31G criteria, the flow stress is 110% of the *specified minimum yield strength (SMYS)*. For the modified B31G and RStreng calculations, flow stress is SMYS + 10,000 psi (SMYS + 68,948 kPa).

flux cored arc welding (FCAW): *See weld.*

foam pig: A type of utility pig. See *pig.*

Folias factor: A parameter used in the assessment criteria such as *B31G, modified B31G* and *RStreng* to account for the bulging/"lipping" of corrosion anomalies under stress. The Folias factor is responsible for the effect that the axial length of corrosion has on predicted burst pressure. The definition of the Folias factor depends on the assessment criteria being used. For B31G, the Folias factor, M, is given by

$$M = \sqrt{1 + 0.8 \frac{l^2}{Dt}}, \qquad \text{if } \frac{l^2}{Dt} \leq 20$$

where l = axial length of the corrosion anomaly
D = diameter of the pipeline
t = nominal wall thickness of pipe

It should be noted that if $\frac{l^2}{Dt} > 20$, the Folias factor is not applied in B31G because the axial length of the corrosion does not result in significant bulging of corrosion anomalies under stress.

For modified B31G and RStreng, the Folias factor is given by

$$M = \begin{cases} \sqrt{1 + 0.6275 \frac{l^2}{Dt} - 0.003375 \frac{l^2}{Dt}}, & \text{if } \frac{l^2}{Dt} \leq 50 \\ 0.032 \frac{l^2}{Dt} + 3.3 & , & \text{if } \frac{l^2}{Dt} > 50 \end{cases}$$

Figure 41 compares the Folias factor for B31G and modified B31G.

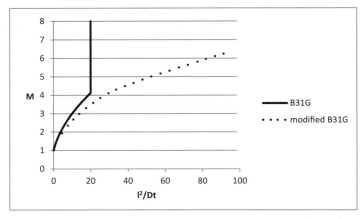

Figure 41. Comparison of the Folias factor for B31G and modified B31G.

forced drainage: See *drainage*.

foreign structure: This term is most often used in the context of cathodic protection systems. Specifically, it is any structure owned by another company or third party that is not intended to be impacted (either positively or negatively) by the cathodic protection system. The nature of the interaction of the foreign structure and the cathodic protection system will be determined by the structure's location relative to the electrical field(s) generated by the cathodic protection system as well the electrochemical properties of the structure itself.

fractography: The study of the surface of fracture surfaces – especially in metals, with specific reference to photography and visual analysis of the fracture surface. Macrofractography involves photographs at low magnification <25×), whereas microfractography refers to photographs taken at high magnification (>25×). A scanning electron microscope (SEM) is often used.

fracture control: This term encompasses several techniques that are used to reduce the probability of a running fracture in a pipeline – this is primarily an issue for larger diameter pipelines (i.e., > NPS 20). The primary approach in fracture control is to design the pipeline to avoid brittle fractures and then ensure that the pipeline has sufficient *toughness* to withstand a ductile fracture. As such, codes typically define *toughness* levels for pipe material as part of minimum design requirements.

The likelihood of a running fracture is determined by the difference in the speed of the decompression wave and the speed of the fracture velocity in the pipe. Running fractures are generated when the velocity of the facture in the pipe wall exceeds the velocity of the decompression wave in the product. Fracture propagation will occur when the fracture velocity and gas decompression velocity curves intersect – any crack will arrest when the two curves become tangential. See Figure 42.

The impact of key parameters on gas decompression velocity and fracture velocity appear in Tables 31 and 32. Thus, the primary means of establishing fracture control is through the use of crack arrestors that decrease the pipe stress either through reinforcement of the pipe wall (e.g., use of a reinforcement sleeve) or an increase in strength (e.g., insertion of heavier wall or higher strength pipe joints at regular intervals).

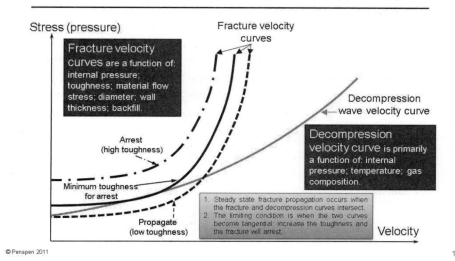

Figure 42. Fracture and gas decompression velocity as a function of pipeline pressure [adapted from (8), courtesy P. Hopkins].

Table 31. Impact of various parameters on fracture velocity (8)

Parameter	Fracture velocity
Diameter	• Larger diameter leads to higher stress levels (for a given toughness and constant pressure) and therefore fracture velocity increases.
Wall thickness	• Lower wall thickness leads to higher stress levels (for a given toughness and constant pressure) and therefore fracture velocity increases.
Internal pressure	• Higher internal pressure leads to higher stress levels (for a given toughness and constant pressure) and therefore fracture velocity increases.
Toughness	• Higher toughness reduces fracture velocity.
Backfill	• The presence of backfill reduces fracture velocity; if the backfill (or soil) is frozen, the fracture velocity is reduced further.

Table 32. Impact of various parameters on gas decompression velocity (8)

Parameter	Effect on fracture velocity (with all other parameters held constant)
Diameter	• Larger diameter leads to higher stress levels (for a given toughness) and therefore fracture velocity increases.
Wall thickness	• Lower wall thickness leads to higher stress levels (for a given toughness) and therefore fracture velocity increases.
Internal pressure	• Higher internal pressure leads to higher stress levels (for a given toughness) and therefore fracture velocity increases.
Toughness	• Higher toughness reduces fracture velocity.
Backfill	• The presence of backfill reduces fracture velocity; if the backfill (or soil) is frozen, the fracture velocity is reduced further.

fracture mechanics: This is the group of theories and models used to describe the behavior of defects in structures. It provides a quantitative method for evaluating structural integrity in terms of applied stress, crack size, and specimen or machine component geometry.

fracture toughness: A generic term for a group of measures that describe the resistance of a material to the extension of a crack. The most common parameters for quantifying fracture toughness include the J integral; the stress intensity factor, K; and the crack-tip opening displacement (CTOD).

free corrosion potential: See *corrosion potential*.

Free-swimming: Describes a pig (either utility pig or smart tool) that is driven by the product flow itself – it is not self-propelled or tethered. The vast majority of pigs used in pipeline applications are free-swimming.

frequency: Describe oscillations or repetitive events. The frequency is the number of complete cycles per unit time. The time between complete cycles is called the period, P; the frequency, f, is the inverse of the period:

$$f = \frac{1}{P}$$

The SI unit for frequency is a hertz (Hz); it is the reciprocal of a second (s^{-1}). The calculation of frequency is relevant to pressure (stress) cycling that could lead to failure through fatigue-based mechanisms.

frost heave: The phenomenon of (primarily) vertical displacement that soils and pipeline experience (primarily) in areas of permafrost – particularly at soil transition points. The exact location and amount of displacement are a function of the rate of heat loss, soil composition and structure (specifically porosity, which potentially facilitates capillary action moving water towards the area of heat loss). The resulting pipeline strains can be significant, and terrains, where frost heave is a risk, require special design considerations and construction practices.

FSM: See *electrical field signature method.*

full encirclement sleeve: See *sleeve.*

fully turbulent flow: See *turbulent flow.*

fungi: One of the most significant contributors to microbiologically influenced corrosion. Fungi can be either single-celled or multi-celled organisms that do not contain chlorophyll. Both "rigid" and "slime" forms of fungi exist with most being non-motile. There are two main categories of fungi: molds and yeasts; however, molds are multi-celled and thought to be more important to corrosion.

Fungi contribute to corrosion principally by two mechanisms: the production of corrosive products (primarily the excretion of organic acids as a by-product of metabolism), and the trapping of material that causes fouling and corrosion. Fungi are can also cause the degradation of other material such as paint, coatings, and oil products.

fusion bond epoxy: See *FBE* and *coating.*

G

galvanic anode: The electrode in a cathodic protection circuit where oxidation occurs – that is, it provides sacrificial protection to the pipeline in question. The term is specifically used to reference the anode in a sacrificial cathodic protection system (vs. an impressed current system). See also *galvanic cathodic protection*.

galvanic cell: See *electrochemical cell*.

galvanic corrosion: Refers to the metal loss that results when two dissimilar metals are in electrical contact, in an electrolyte, such that one of the metals is oxidized, or corroded, due to the difference in their relative positions in the galvanic series (i.e., that is one metal is less noble than the other). The two metals then naturally form what is termed a *galvanic couple* where a (galvanic) current flows between the metals, due to their differing affinities for electrons, causing the less noble metal to corrode. On pipelines, there are several scenarios that can result in galvanic corrosion, including oxidation of the metal within the heat affected zone of a weld, at casings, repair sites and other locations where appurtenances have been installed and the differing metals remain in electrical contact. Should galvanic corrosion be a concern, dielectric (i.e., non-conductive) fittings can be used between dissimilar metals to ensure there is not electrical contact.

On pipelines, cathodic protection systems make use of a galvanic pair to protect the steel from corrosion. The steel pipeline is connected to a sacrificial anode, usually made of zinc or magnesium, to create a galvanic couple. Thus, the current flows from the steel to the sacrificial anode, which is then preferentially corroded.

galvanic couple: See *galvanic corrosion*.

galvanic cathodic protection: A form of cathodic protection that uses only the inherent differences between the material properties of two or more metals to protect a structure from corrosion. In contrast, impressed current cathodic protection systems apply current to the structure and the anodes from an external power supply. See also *cathodic protection*.

galvanic series: A list of metals and alloys arranged in order according to their corrosion potentials in a given environment. Typically, noble metals are listed at the top of the list and active metals at the bottom of the list. A noble metal or alloy in electrical contact with a more active metal or alloy will cause the more active to material to corrode by galvanic corrosion. See also: *electromotive force series*.

Gamma distribution: A probability distribution defined for positive values. Mathematically, the probability density function of the gamma distribution, $G(x)$, is defined as

$$G(x) = \begin{cases} \dfrac{1}{\Gamma(\alpha)\beta^{\alpha}} x^{\alpha-1} e^{-x/\beta}, & 0 < x < \infty \\ \\ 0, & x \le 0 \end{cases}$$

where α and β are the shape parameters of the distribution; $\Gamma(\alpha)$ is the gamma function:

$$\Gamma(\alpha) = \int_{0}^{\infty} \left(\frac{x}{\beta} \right)^{\alpha-1} e^{-x/\beta} \frac{1}{\beta} dx.$$

Table 33 shows selected properties of the gamma distribution. The exponential and chi-squared distributions are special cases of the gamma distribution. See *exponential distribution* and *chi-squared distribution*. See also *probability distribution*. A sample of the gamma distribution with various parameters is shown in Figure 43.

Table 33. Properties of gamma distributions

Parameters	α and β
Domain	$0 \le x \le \infty$
Mean	$\alpha\beta$
Variance	$\alpha\beta^2$
Mode	$(\alpha-1)\beta$

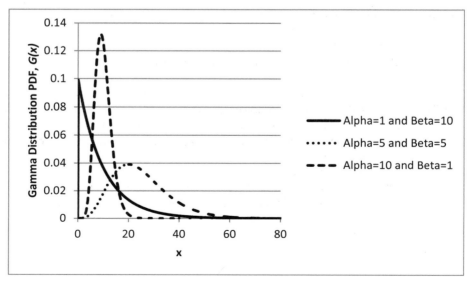

Figure 43. Gamma distribution for various values of α and β. The special case of α=1 is the exponential distribution.

gas: A fluid and one of the three principal phases of matter: solid and liquid are the other two phases. A gas must be confined on all sides to be contained. As a pipeline product, the term "gas" usually means natural gas, which is a hazardous material, inflammable, toxic, and often corrosive. Not surprisingly, the composition of natural gas varies widely between production and transmission systems with transmission systems having relatively strict criteria regarding gas quality – especially as it relates to moisture, CO_2, and H_2S content. Table 34 provides a general guide to the composition of most natural gas mixtures. The ramifications of composition in pipeline integrity, affect factors such as the corrosivity, flow regime, running facture propagation, potential impact radius from the ignition, and toxicity.

Table 34. Indication of the composition of natural gas mixtures (8)

Element		Concentration
Methane	CH_4	70 to 98%
Ethane	C_2H_6	1-10%
Propane	C_3H_8	Trace – 5%
Butane	C_4H_{10}	Trace – 2%
Pentane	C_5H_{12}	Trace – 1%
Hexane	C_6H_{14}	Trace – 0.5%
Longer-chain hydrocarbons	C_{7+}	0% to Trace
Nitrogen	N_2	Trace – 15%
Carbon Dioxide	CO_2	Trace – 1%
Hydrogen Sulfide	H_2S	Trace – 90% (?)

gas line: A pipeline (flow, gathering, transmission, distribution, or service line) used to transport chemical product in a gaseous form, usually natural gas.

gas metal arc welding (GMAW): See *weld*.

Gas Research Institute (GRI): See *Gas Technology Institute*.

Gas Technology Institute (GTI): A not-for-profit research and development organization. Its goal is to provide economic value to the energy industry and its customers, while supporting government in achieving policy objectives. GTI has three primary objectives, which span the energy industry value chain:
- Expanding the supply of affordable energy
- Ensuring a safe and reliable energy delivery infrastructure
- Promoting the efficient use of energy resources

gas tungsten arc welding (GTAW): See *weld*.

gate valve: A valve with a closing disk element that is moved across the stream, usually in a groove or slot for support against pressure. Often, the closing element may be the shape of a wedge (not a disk), in this case the valve is referred to as a "wedge" type gate valve. Gate valves are usually used to get a straight line of flow with a minimum restriction – as such they are not suited for flow control but rather full shut off or flow. Depending on the specific design, gate valves can be problematic for pigging operations. Figure 44 shows the basic form of a gate valve, and Table 35 cites major subtypes.

gathering line: A system that usually encompasses the pipelines that gather product, usually petroleum or natural gas, from the production facilities and transport it to a central point for processing, refining, and/or further transmission. These lines tend to be small diameter and largely operate at relatively low pressures. Other characteristics of distribution lines are that they are primarily constructed within dedicated rights-of-way in Class 1 locations. Further, in the case of gas pipelines, the product is not odorized and the primary failure mode is leak, because the pipelines most often operate at significantly lower stress levels.

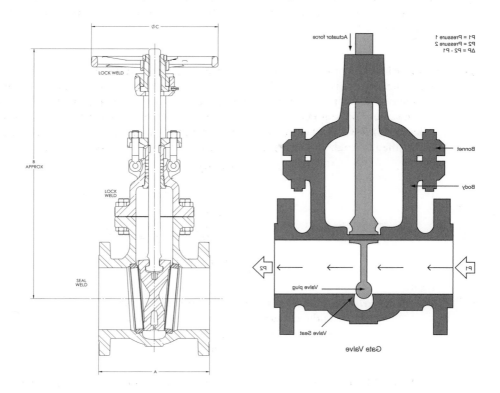

Figure 44. Gate valve schematic (a) and operation (b). (200)

Table 35. Major subtypes of gate valves (27)

Subtype	Characteristics
Rising	Stem rises as valve is opened and lowers as valve is closed – providing a clear visual indication of valve position
Non-rising stern	The stem remains vertically stationary – some sort of pointer is then used to indicate valve position

As such, the pipeline integrity programs associated with gathering systems can be quite different than those that would be considered appropriate for large diameter transmission or distribution systems. The key differences such as product quality, diameter, and consequence of failure can drive vastly different decisions in the area of inhibitors, inspection programs, and repair vs. replace decisions. See also *pipeline*.

gauging pig: A utility pig used to detect restrictions in a pipeline. Specifically, a utility pig is mounted with a slotted metal plate, or plates, of a specified diameter less than the minimum internal diameter of the pipeline. Pipe bore restrictions less than the plate diameter or short radius bends will permanently deflect the plates. The diameter of the smallest plate to be deflected gives the size of the maximum restriction on the pipeline. See *pig*.

gel pig: A pig train composed of several utility pigs and semi-solid (i.e., cross-linked) water, solvent, or other chemicals usually for the purposes of batch separation, cleaning, or inhibitor delivery. See pig.

general corrosion: 1. A form of deterioration that is distributed more or less uniformly over a surface. The distribution of external general corrosion is dependent on coating disbondment and tends to be more common on certain parts of the pipeline; for example, it may be more common on the lower half of the pipeline (from the 3 to 9 o'clock positions). **2.** General corrosion refers to the approximate shape of metal loss defect as identified through in-line inspection. Specifically, the dimensions of the anomaly are

$$length < 3A$$

and

$$width < 3A$$

$$A = \begin{cases} t, \ if \ t < 10\,mm \\ 10\,mm, \ if \ t \geq 10\,mm \end{cases}$$

where

$length$ is the axial length of the corrosion anomaly;
$width$ is the circumferential width of the corrosion anomaly;
t is the wall thickness; and
A is the wall thickness in mm.

The definition is based on a document published by the *Pipeline Operators Forum*; additional detail is provided in *defect classification*.

geographic information system (GIS): A computerized system that combines cartographic features and databases to overlay reference information (such as corrosion defects, extent of soil erosion, pipe deformation, etc.) on computer-generated maps. GIS comprises the set of data, processes, software, and hardware used to store, process, analyze, and display data that has a spatial component (i.e., tied to a geographic location). This is a relatively new field that has developed with the advent of faster and cheaper computing power. There does not appear to be a standard definition of regarding the main components of a GIS system; however, broadly speaking the main elements would include the spatially referenced data, the data visualization/user interface, computer hardware, and software.

geomatics: A relatively new discipline that deals with information that is referenced spatially (i.e., via location). The field has emerged with the advent of readily available, highly accurate, and relatively inexpensive location information such as the global positioning system. The discipline requires a merging of skills in several key areas cited in Table 36.

Table 36. Key skills required in geomatics engineering (138)

Area	Key skills
Data acquisition	• Understand and use terrestrial, marine, airborne, and satellite-based systems to acquire spatial and other data.
Modelling	• Mathematically transform spatially referenced data from different sources into common information systems with well-defined accuracy characteristics based on the scientific framework of geodesy.
Analysis	• Undertake positioning and navigation calculations. • Extract information from digital imaging. • Undertake mapping based on air photos or remote sensing
Data Management	• Work with Land and Tenure Systems as well as Geographical Information Systems.

geometry tool: A type of smart pig used to assess the nature of the pipeline geometry. This type of inspection is often conducted prior to other inspections for corrosion or cracking; however, these tools are also used to inspect the line for dents, wrinkles, and other mechanical damage. They are often combined with inertial mapping capabilities for mapping and to detect movement of the pipe due to *geotechnical threats*. The term is used interchangeably with *deformation tool* and *caliper tool*. See *pig* for a detailed discussion of this class of tools.

geotechnical threat: The term refers to a group of factors, based on the behavior of soil and rocks that give rise to a greater probability of damage to the structural integrity of the pipeline. Table 37 shows the most common forms of geotechnical threats.

Table 37. Common forms of geotechnical threats on a pipeline (26)

Type	Description
Erosion	• The breakdown and subsequent movement of rock, soil, backfill or other natural material due to the action of gravity, ice, wind or water on the surface of the earth • See *erosion*
Frost heave / thaw settlement	• The phenomenon of (primarily) vertical displacement that soils and pipeline experience (primarily) in areas of permafrost – particularly at soil transition points • See *frost heave* and *thaw settlement*
Seismic activity	• A threat to a pipeline either directly or indirectly. Ground movement may cause deformation and possible failure of the pipeline. Indirect threats include landslides for onshore pipeline and turbidity currents for offshore pipelines.
Slope instability	• Can cause strain in a pipeline and lead to its eventual failure.
Subsidence	• The overall reduction in the height of ground level over time due to geological effects is termed subsidence. • See *subsidence*

girth weld: A circumferential weld joining two joints of pipe. See Figure 45.

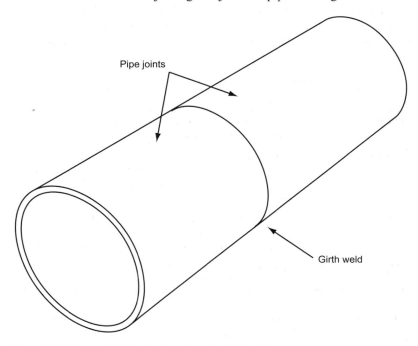

Pipe joints

Girth weld

Figure 45. Schematic indicating location of girth weld.

girth weld number: A unique identifier assigned to each girth weld on a pipeline. Because girth welds are easily identifiably by both the tool and in the field, as well as widely and relatively evenly distributed on a pipeline, they make good reference locations. Typically, girth welds are numbered by 10's: for example 100, 110, 120, 130, etc. The reason for this numbering sequence is to allow for repairs and other events to occur on the pipeline and still keep the girth numbers sequential. After a cut-out repair, the girth-weld numbers from the removed girth welds are not reused. The new girth welds are numbered using intermediate numbers to maintain the sequential girth-weld numbering. Figure 24 illustrates the numbering of girth welds before and after a repair.

GIS: See *geographical information system.*

global positioning system (GPS): A navigational system using satellite technology to provide a user an exact position on the earth's surface. A GPS receiver calculates its location from data it receives from orbiting satellites. The satellites are equipped with atomic clocks and continually transmit their location and precise time information. The GPS receiver receives information from four or more satellites. It then calculates its position by precisely timing the signals sent by the GPS satellites. The receiver calculates the transit time of each message and computes the distance to each satellite. Geometric trilateration is used to combine these distances with the location of the satellites to determine the receiver's location. Speed and direction of travel are derived from the changing position of the GPS receiver.

globe valve: Probably the most common type of valve across all applications, the name comes from the globular shape of the valve body. The closing element is a spherical, flat, or rounded gasket, which is moved into or onto a round port. This type of valve is suited both to throttling and general flow control. A schematic illustrating the basic form of a globe valve is shown in Figure 46.

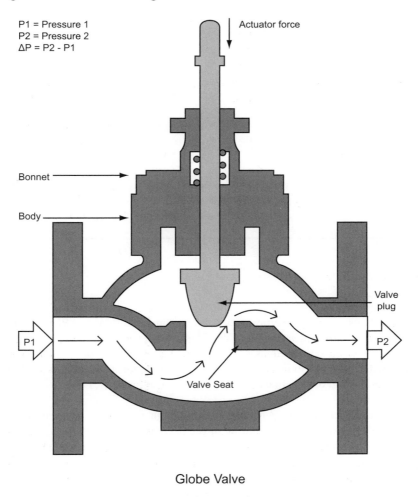

Globe Valve

Figure 46. Principle of globe valve operation.

GMAW: Refers to Gas Metal Arc Welding. See *weld*.

gouge: A form of mechanical damage, where part of the pipe wall thickness is physically removed resulting in a reduction of the pipe's load carrying capability. Some of the resulting concerns stem from damaged coating (increased likelihood of corrosion and cracking), highly localized strain-hardening (i.e., increased likelihood of cracking in cold worked area) and existence of a stress concentration (due to the geometry of the gouge and potentially associated deformation). One of the most common causes of gouging on pipelines is excavation equipment scraping the pipeline surface.

Assessing gouges is a relatively complex process that requires the consideration of a number of factors, some of which are shown in Table 38.

Table 38. Considerations when assessing gouges (8)

Factor	Description
Pipe properties	• To undertake assessment, ensure pipe material toughness >20J
Associated concerns	• Denting may exist in association with gouge • Cracking may exist in association with gouge • Coating damage • Due to the increased threat from cracking (e.g., fatigue or environmentally assisted cracking), need to consider additional conservatism in analysis
Standards	• ASME B31.8 (2003) requires all gouges to be removed • Australian Standard AS 2885.3 states grooves, notches <0.25mm can be considered "harmless" • If >0.25mm but ≤10%wall thickness, can be removed by grinding • Wall thickness loss due to grinding may be assessed in traditional manner
Assessment	• Equations developed by Battelle • Use BS 7910/API 579 to assess for fatigue crack growth, if pipeline is subjected to cyclic stresses • May need to "increase" defect depth to account for strain hardened layer in pipe material

GPS: See *global positioning system*.

grade of pipe: Refers to pipe properties as defined by standards such as API Spec 5L standard for seamless and welded line pipe. The 5L standard was originally issued in 1920, but has been revised several times – perhaps most significantly, the standard was harmonized with ISO 3183 to be consistent with international standards. Table 39 lists common grades used on pipelines and their mechanical properties.

grain: An individual "crystal" in a metal or alloy, about the size of a fraction of an inch, in which the atoms are arranged in an orderly pattern. These grains, or crystals, are separated by grain boundaries, which are narrow zones in a metal where the atomic structure transitions from one crystallographic orientation to another. While the grains have a high degree of atomic organization, the intergranular zones are more disorganized. In pipeline integrity, specific mechanisms of cracking (e.g., transgranular vs. *intergranular*) and corrosion are influenced by grains (and grain boundaries), and this can be an important descriptive parameter that allows more accurate identification of the specific nature of the defect.

graphite anode: Solid rods of graphite with a small amount (0.2%) of ash and are one of the most common types of anodes used with *impressed current cathodic protection* systems.

grid: Generally, a framework of squares or lines arranged in a regular pattern on a surface to specify locations on that surface. In the context of pipeline integrity, grids are often used during an excavation as an essential part documenting the depths of the corrosion that is found. After cleaning the pipe surface, the grid is drawn on the outer wall of the pipe. Then NDT measurements are made in each square. The grid assists in being able to specify the location of the NDT measurements and to assist the technician to systematically assess an area of pipe wall with the appropriate burst pressure equation.

grinding: A repair method for some shallow (typically ≤10% - 12.5% of wall thickness) defects or imperfections on a pipeline. The repair involves the removal of the metal from the pipe surface – along with the flaw – to remove any stress concentration and hardened material associated with the defect. The metal is removed through hand filing or power grinding. Grinding does result in a thinner wall in the vicinity of the repair, and care is required to ensure that the remaining metal is sufficient for the integrity of the pipeline. Examples of anomalies that are typically addressed in this manner include shallow surface cracks, gouges, and arc burns.

The technique can also be used to assess the depth of cracks where several areas of a colony, felt to be indicative of the maximum depth within the colony, may be subjected to a grinding procedure. To assess the depth of cracking, a predetermined amount of metal is removed from the surface of the pipe by grinding. The surface is then re-examined for cracks. If any cracks are visible, then another layer of metal is removed, and the process is repeated until no cracks are visible. The amount of metal removed is used to determine the depth of the cracking. Figure 47 illustrates the metal removed to grind out a crack.

Table 39. Mechanical properties of common pipe grades (20)

Pipe Grade USC / SI	Yield Strength (ksi)		Tensile Strength (ksi)		Yield Strength (MPa)		Tensile Strength (MPa)	
	Min	Max	Min	Max	Min	Max	Min	Max
L245 / A25	35.5	65.3	60.2	110.2	244.8	450.2	415.1	759.8
L360 / X52	52.2	76.9	66.7	110.2	359.9	530.2	459.9	759.8
L415 / X60	60.2	81.9	75.4	110.2	415.1	564.7	519.9	759.8
L555 / X80	80.5	102.3	60.6	119.7	555.0	705.3	417.8	825.3
L690 / X100	100.1	121.8	110.2	143.6	690.2	839.8	759.8	990.1
L830 / X120	120.4	152.3	132.7	166.1	830.1	1050.1	914.9	1145.2

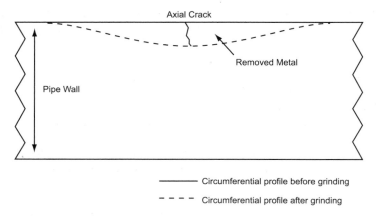

Figure 47. Example of a grind repair [adapted from (8)].

grit blasting: A method of cleaning the surface of a pipeline. The method blasts small particles (grit) at a high velocity at the surface of the pipeline to clean away any material stuck to the surface. The grit particles are often made of iron, steel, or sand.

There are three principal reasons to grit blast a pipeline. The application of coating requires a clean surface to adhere. Grit blasting can also reduce the susceptibility of the steel to SCC cracking by introducing a compressive stress to the surface of the steel. In an investigative excavation the surface of the pipeline must be clean for the accurate use of NDE methods, such as an ultrasonic probe (UT) and magnetic particle inspection (MPI). However, in some cases, aggressive grit blasting can plastically deform the surface of the steel and obscure the existence of cracking for MPI. Various grits have been used to reduce damaging of the pipe surface prior to a MPI, e.g., walnut shells.

grooving: A type of metal loss feature resembling a long narrow defect that is either axially or circumferentially aligned. Axial grooves are long and narrow, but wider than axial slotting. Circumferential grooves are wide and short but longer than circumferential slotting. See *defect classification* for a detailed regarding the range of dimensions for this defect type.

ground bed: A grouping of anodes (i.e., usually more than one anode) installed below the earth's surface for the purpose of supplying cathodic protection to pipeline or other metallic structure. Ground beds can be arranged in a variety of configurations. Some more common configurations include single vertical anode chain, single horizontal anode chain, parallel array of vertical anodes, and parallel array of horizontal anodes.

ground fault current: See *fault current.*

group: A collection of metal-loss clusters (usually as identified by an in-line inspection tool) that are close to one another and need to be treated as a unit to calculate burst pressure. Whereas *clusters* are collections of individual box anomalies, groups are collections of clusters. Like clusters, groups are identified by the application of an *interaction rule.*

growth assay: One of the most common methods of identifying the potential for microbial-influenced corrosion (MIC). The assay is a kit used to determine if certain bacteria types are present by attempting to culture them in pre-prepared medium.

GTAW: Refers to gas tungsten arc welding. See *welding*.

guided wave UT (GWUT): Guided-wave UT is a long-range ultrasonic method to investigate potential metal loss on a pipeline. The typical application of guided-wave UT is to investigate the condition of a pipeline from a bell-hole excavation for some distance beyond the excavation. In some cases, the bell-hole excavation dug adjacent to an excavation location would be cost prohibitive.

The method uses a device, called a collar, which straps around the pipeline and induces ultrasonic waves in the pipeline. The collars can often generate a variety of wave modes in the pipeline. The guided waves travel along the length of the pipeline and are reflected back by any deviation from the normal pipe, such as girth weld and corrosion. The reflected signals are recorded at the excavation site. The time-of-flight is used to calculate the location (distance up or downstream) of the anomalies.

The distance that can be investigated by the guided-wave UT depends on the amount of attenuation of the wave down the length of the pipeline. Attenuation is dependent on the type of coating, the type of soil, and the fluid content of the pipeline (i.e., effective transmission of ultrasonic signals in gases is practically difficult – liquids are much more efficient transmitters of such signals). In the ideal case, the investigation can be hundreds of meters. However, sometimes the signal is attenuated in only a few meters. See also *ultrasonic testing*.

Gumbel distribution: One of the principle distributions used in statistics of extremes; it has been used to estimate the distribution of corrosion pit depths on a surface. The PDF, $f(x)$, of the Gumbel distribution is given by

$$gb(x) = \frac{\exp(-(x-\mu)/\beta)\exp(-\exp(-(x-\mu)/\beta)))}{\beta}$$

and the CDF, $Gb(x)$, is given by

$$Gb(x) = \exp(-\exp(-(x-\mu)/\beta))$$

where μ = location parameter of the distribution
β = shape parameter

Table 40 shows selected properties of the Gumbel distribution. Figure 48 shows the shape of the Gumbel distribution for various values of μ and β.

GWD: See *girth weld*.

Table 40. Gumbel distribution parameters

Gumbel distribution properties	
Parameters	μ and β
Domain	$-\infty \le x \le \infty$
Mean	$\mu + \gamma\beta$
Variance	$\dfrac{\pi^2}{6}\beta^2$
Mode	μ

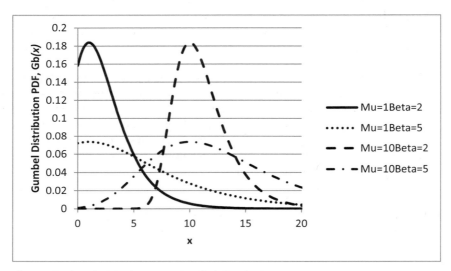

Figure 48. Plot of PDF when using Gumbel distribution.

H

H₂S: See *hydrogen sulfide*.

half-cell: Half of an electrochemical reaction. It can be either the anodic or cathodic part of an electrochemical reaction. In an electrochemical reaction the anodic half of the reaction produces electrons, while the cathodic half of the reaction consumes them. See anodic reaction and cathodic reaction. The half-cell, also referred to as the reference half-cell, is used as a reference for the vast majority of measurements needed as part of installing, monitoring, and maintaining a cathodic protection system.

Hall effect sensors: Electronic sensors that measure the magnetic flux using the Hall effect; i.e., they are calibrated to detect and measure the magnetic flux perpendicular to the sensor based on the potential difference induced by the flow of current.

Hall-effect sensors are being used increasingly on MFL tools as they have the advantage of being smaller than earlier generation coil sensors. Further, Hall effect sensors measure the absolute magnetic field, whereas coil sensors measure changes in magnetic flux.

The sensors themselves are named for E. H. Hall who in 1879 devised an experiment to show that the negatively charged motion of electrons were responsible for the flow of electric current in a metal. The effect is illustrated in Figure 49.

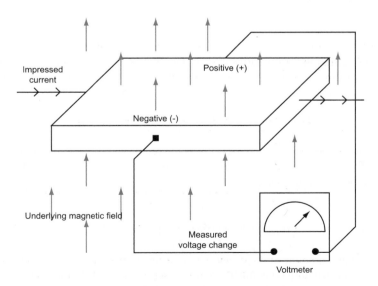

Figure 49. Illustration of the Hall effect used for MFL sensors.

halo: The signature signal produced by a change in material properties when the signals of two successive MFL runs, undertaken at two different magnetization levels, are compared. Specifically, the term is used in reference to a method used to detect hard spot anomalies and mechanical damage resulting in a hardening of the steel. The method involves the running of two MFL inspections at different magnetization levels. The first MFL inspection is done as per standard operations where the pipe wall is fully saturated. At the high magnetization level, the tool is relatively insensitive to changes to the magnetic properties of the steel. However, for the second inspection, the magnets are removed from the tool and readings rely on the residual magnetization in the pipe wall. At the lower magnetization level, the tool is sensitive to changes in material properties. Thus, any differences in the signals from the two inspections will be indicative of material property changes such as dents or hard spots. The term is based on the fact that the difference in the signals resembles a "halo" around the dent or hard spot.

hand-tool cleaning: This is the removal of loose rust, mill scale, paint, and other debris from the pipe surface to the degree specified, by hand chipping, scraping, sanding, and wire brushing.

hard spot: An area of steel that has a higher-than-normal hardness. This hardness is due to a local difference in metallurgical properties. Table 41 lists the common causes of hard spots. Hard spots are not desirable on a pipeline because they are susceptible to failure from various mechanisms – most commonly they are susceptible to cracking due to greater potential for hydrogen embrittlement. Because they do not bend as easily as normal steel, they often cause flat spots on the surface of the pipe where the radius of curvature is much greater than the rest of the pipe, resulting in local stress concentrations. Hard spots are often detectable through the use of magnetic flux leakage tools. See *halo*.

hardener: See *curing agent*.

hardness: The measure of the resistance of a material to plastic deformation by indentation. Figure 50 compares the most common methods to measure hardness, which are the Brinnell, Vickers, and Rockwell C hardness tests. Also see Table 42.

Table 41. Common causes of hard spots in pipelines. (8)

Cause	Description
Inappropriate quenching	• Accidental quenching of the plate used to form the pipe can result in hard spots • Specific vintages of pipe are known to have hard spots due to problems during the fabrication process
Cold work	• Localized hardening can occur as a result of mechanical damage (e.g., dent, gouge, or field bending)
Welding damage	• Improper control during the welding process can lead to localized hardening in the vicinity of arc burns and within the heat affected zone (HAZ)

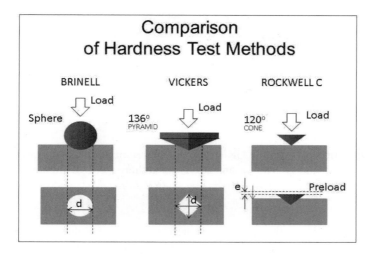

Figure 50. Comparison of Brinnell, Vickers, and Rockwell C hardness tests [adapted from (195)].

Table 42. Equivalence of Brinnell, Vickers, and Rockwell C hardness tests [adapted from (195)]

Brinnell (3,000 kg)	Vickers (100 kg)	Rockwell C (150 kg)
388	410	41.8
363	383	39.1
341	360	36.6
321	339	34.3
302	319	32.1
285	301	29.9
262	276	26.6
241	253	22.8
237	248	22.0
229	241	20.5

HAZ: See *heat affected zone*.

HAZ corrosion: This is the preferential corrosion of the heat-affected zone around a weld, such as a seam weld or girth weld. The heat of welding causes a metallurgical change in the steel, which can cause the heat-affected zone to be anodic to the rest of the pipe. The change creates a local galvanic cell, which causes the heat-affected zone to corrode.

hazard: Anything that may cause a failure or otherwise threatens the integrity of a pipeline or facility. The term is used interchangeably with threat. See threat.

hazardous liquid: Any liquid that poses a threat to human and animal life, property, or the environment if released. Most pipeline products are hazardous as they are usually inflammable or toxic liquids: such as crude oil, petroleum products or carbon dioxide (carbon dioxide is usually in a dense or liquid state at pipeline pressures). It should be noted that many seemingly innocuous materials can be considered hazardous if released into the environment in large enough quantities (e.g., carbon dioxide).

HCA: See *high consequence area*.

Health and Safety Executive (UK): The UK's independent regulator overseeing work-related health, safety and illness. The organization has authored one of the key references documents in the area of assessing risk. See *ALARP*. (194)

heat affected zone: The portion of the base metal (not the weld material) that is not melted during brazing, cutting, or welding, but whose microstructure and properties are altered by heat from these processes. Figure 51 shows a schematic cross-section of a HAZ.

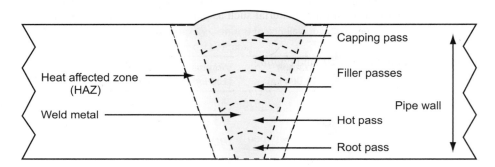

Figure 51. Cross-section of a weld showing the heat-affected zone (HAZ).

heat number: An identification number that identifies the batch from which the steel was poured. In steel making, a heat is the batch of steel made from a single melting operation; thus, all pipe joints from a single heat have the same chemical composition. Sometimes, but not universally, the first digit of the heat number indicates the furnace number; the second digit indicates the year in which the heat was melted. The last three (and sometimes four) digits show the number of batches of metal melted in the furnace in that year.

heat treatment: The heating and cooling a solid metal or alloy in such a way as to obtain desired properties – specifically, the rate of heating, hold, and cooling is controlled to yield a specific grain size and (metallic) phase that has the desired properties. The more common heat treatments used for steels include annealing, quenching, interrupted quenching, austempering, and tempering. It should be noted that heating for the sole purpose of hot working is not considered heat treatment.

heavy-wall pipe: Pipe with a wall thickness greater than that required for the given, pressure, diameter, and operating conditions in a Class I location. However, in practice, the term is used largely to differentiate lengths of pipe where the wall thickness is greater than most pipe directly upstream or downstream at a given location. For example, heavy-wall pipe is used at road and rail crossings, river crossings, and higher Class locations.

HIC: Also referred to as *hydrogen induced cracking*. See *hydrogen damage*.

High Consequence Area (HCA)[1]: Generally, an area in which a pipeline failure would have a potentially high impact on human health, life and property and, as such, the definitions differ for liquid and gas pipelines. The term arises from US regulation where the definition for liquid pipelines appears in Part 195.450, and the definition for gas is and is defined in 49 CFR Part 192.761.

high pH SCC: See stress corrosion cracking.

high-pressure water jetting: The term is used interchangeably with high-pressure water cleaning. See water jetting.

high-resolution in-line inspection: A term usually applied to modern magnetic-flux-leakage tools to differentiate them from older standard-resolution (SR) tools (also called conventional-resolution or low-resolution tools). The basic technology is the same for both high-resolution and standard-resolution MFL; however, incremental improvements in several components have made the more modern high-resolution tools more accurate. The difference between high-resolution and standard-resolution MFL was once a topic of debate among providers. More recently, almost all in-line inspection providers offer only high-resolution tools and the difference has become moot. Because tools vary from vendor to vendor, there is no definitive list of differences; however, key variations between high-resolution and standard-resolution tools are shown in Table 43.

high-silicon cast iron anode: Anodes used in impressed current cathodic protection (CP) systems. These anodes contain 14-15% silicon and other alloying elements. The principal advantage of the high-silicon anodes over ordinary cast-iron anodes is the consumption rate. Ordinary cast-iron anodes lose about 9kg/amp-yr (20 lb/ amp-yr), whereas high-silicon cast iron anodes lose only 0.05-0.11kg/amp-yr (0.1-0.25 lb/amp-yr).

high-strength steel: While there is no definitive threshold above which steels are considered to be high strength, in the transmission pipeline sector, those materials that are X70 or above, are generally considered to be in this category. Specifically, the advent of these materials has required advances in welding techniques, and they represent a practical delineation point for classifying materials based on strength.

---

1 The information associated with regulation and regulatory guidelines has been prepared for information purposes only. The material is not intended to be, nor does it constitute, a legal interpretation of the regulation or associated codes. As such, the reader should ensure that appropriate and up to date advice, specific to their situation is obtained when necessary.

Table 43. Key differences between standard and high-resolution corrosion pigs (124)

Characteristic	Standard resolution	High resolution
Sensors	• Coil type sensors – measuring changes in the magnetic field • Sensors axially oriented only • Lower sensor density	• Hall effect type sensors – measuring the magnetic field in absolute terms
Magnetization levels	• Lower magnetization levels	• Full saturation of the pipe wall (unless conducting an inspection to detect changes in material properties)
Defect discrimination	• No/limited discrimination of whether anomaly is on internal or external pipe surface	• Second set of sensors allows internal/ external wall discrimination
Sizing accuracy	• Lower sizing accuracy • Typical depth accuracy is 15% of nominal wall thickness; 80% of the time	• Higher sizing accuracy • Typical depth accuracy is 10% of nominal wall thickness; 80% of the time

high-temperature hydrogen attack: See *hydrogen damage*.

hold point: Refers to the specific pressure level(s) that a pipe section is subjected too during pressure testing. See *pressure testing.*

holiday: A hole or discontinuity in the protective coating on the pipeline surface and can expose the external pipe surface to a potentially corrosive environment.

holiday detection: Devices designed to locate discontinuities in the pipe coating. Most methods are based on detecting the presence of an electrical discontinuity. The two most common methods of holiday detection for excavated (i.e., physically accessible) pipe appear in Table 44. Several methods are available to locate and size coating holidays on buried pipe. They include *close-interval survey*, *direct current voltage gradient* (DCVG), *alternating current voltage gradient* (ACVG), and *Pearson survey*.

hook crack: A type of flaw typically found in the welds of older low-frequency electric-resistance welded (ERW) pipe caused by laminations at the edge of the steel plate. A hook crack is formed when the lamination, which was originally parallel to the surface of the pipe, is pushed into the ERW area and trimmed at the surface. Hook cracks often cause fatigue cracking originating from their tips. Figure 52 shows the formation of a hook crack.

Table 44. Most common methods used to detect coating holidays (195)

Method	Description
Low-voltage wet sponge test	• The method uses a sponge and water to find the holiday where the water provides the electrical continuity through the holiday. • Typical voltages needed are in the 30-90 volt range. • Generally used to locate holidays in thin coatings with a thickness of 0.5mm (20 mil) or less. • Only applicable when pipe is exposed during construction or repair.
High-voltage spark test	• The method uses a soft wire brush, which runs over the coating surface. The presence of a holiday is indicated by a spark between the brush and the pipe all. The method can also detect thin areas of the coating. • Typical voltages for this technique are in the 1,000-40,000 volt range. • Generally used to locate holidays on thicker coatings with a thickness greater than 0.5mm (20 mil). • Only applicable when pipe is exposed during construction or repair.

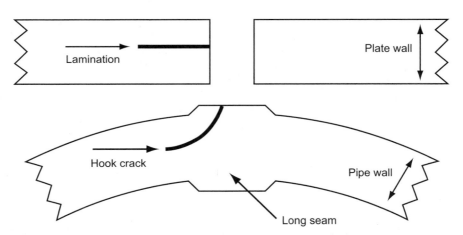

Figure 52. Formation of hook crack from lamination in ERW procedure (195).

hoop stress: See *circumferential stress*.

hot applied coating: Refers to the range of external pipeline coating formulations that require the use of heat for effective application – regardless of the constituent compounds. Thus, this is a generic term often used to describe the numerous coating systems most often used where the coating is mill/factory applied. See coating.

hot bend: The process of introducing curvature into a joint of pipe (i.e., deforming it plastically) at a temperature and strain rate such that recrystallization occurs simultaneously with the deformation, thus avoiding any strain hardening.

hot cut: An operation in which the gas is not fully evacuated from the pipeline before proceeding with repair operations. To eliminate the chance of an explosion, a positive flow of gas is essential and the gas is purposely ignited prior to welding. Very specific procedures are required in a situation to ensure a safe and successful operation.

hot pass: The second layer of weld material laid down during a welding operation. Figure 51 shows a schematic of passes in a weld. See *welding passes.*

hot tap assembly: A customized tee fitting used during hot tap operations (see *hot tap*). A hot tap is one of the steps used in establishing a temporary bypass for a pipeline section while it remains in operation. Specifically, it is the step involving cutting into the pipeline through a customized tee that is either welded or clamped to the pipeline. See Figure 54.

hot tap: One of the steps used in establishing a branch connection for a pipeline section while it remains in operation. Specifically, it is the step involving cutting into the pipeline through a customized tee that is either welded or clamped to the pipeline. See Figures 53 and 54. See *hot tap assembly* for details regarding the customized tee fitting used.

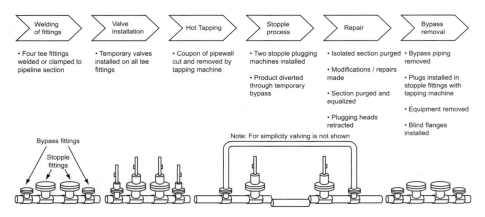

Figure 53. The process for installing a temporary bypass using hot tapping (9).

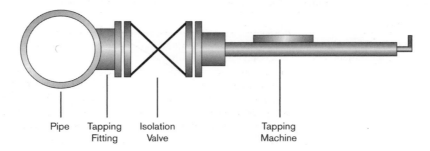

Figure 54. Example of hot tap assembly. (6)

HSE: See *Health and Safety Executive.*

hydrogen blistering: See *hydrogen damage.*

hydrogen cracking: A form of cracking that most often occurs in the heat affected zone of the weld. Generally speaking, there are three conditions, shown in Table 45, that must co-exist in order for this form of hydrogen damage to occur. See *hydrogen damage.* Also known as cold cracking or hydrogen cold cracking.

hydrogen damage: Refers to a broad range of hydrogen-driven mechanisms that can negatively affect the integrity of steel and other metals and alloys. Table 46 shows the two most common mechanisms relevant to transmission pipelines. Other forms of hydrogen damage, less relevant in the context of transmission pipelines, include internal hydrogen precipitation, hydrogen attack, and hydride formation. In terms of hydrogen sources, Table 47 shows four main contributors to the presence of hydrogen:

Table 45. Conditions needed for hydrogen cracking (6)

Factor	Description
Hydrogen in weld	Hydrogen can be introduced into the weld from a number of sources including: • the welding process itself (e.g., the electrode coating, flux) • grease, oil and solvents • moisture (such as from high humidity) • corrosion product (rust is a hydrated oxide) • wet surfaces
Susceptible microstructure	A susceptible microstructure can be produced through the welding process. Specifically, the combination of the metal's chemical composition, weld cooling rate and subsequent thermal treatment (if any) may result in the HAZ having a hard (often martensitic) structure more susceptible to cracking
Stress acting on weld	Additional tensile stresses may be acting on the weld and may be residual stresses (from the construction process for example) or may be imposed later in the life of the pipeline through geotechnical concerns such as pipe settlement

Table 46. Major types of hydrogen damage on pipelines (8)

Mechanism	Description
Hydrogen embrittlement	• Atomic hydrogen easily migrates into the metal and causes a loss of ductility called embrittlement. • The loss of ductility of the metal is due to the presence of dissolved hydrogen in the lattice structure of the metal or alloy. • The loss of ductility increases the pipe's susceptibility to various forms of cracking (terms include: hydrogen induced cracking, hydrogen stress cracking, stress-oriented hydrogen induced cracking – these terms are often used interchangeably).
Hydrogen blistering	• Hydrogen can also accumulate in small traps, or voids in the pipe wall – recombining into molecular hydrogen (atomic hydrogen can easily migrate into steel but molecular hydrogen cannot). • When sufficient hydrogen accumulates, the pressure can cause blistering or tearing of the pipe wall.

Table 47. Sources of hydrogen in pipelines (195)

Mechanism	Description
Corrosion product	• Hydrogen is a by-product of most corrosion reactions
Cathodic protection	• Cathodic protection negatively polarizes the pipeline, causing positively charged hydrogen ions(H+) in the soil to be electrostatically attracted to the pipeline; thus, cathodic protection potentials must be sufficiently high to protect the pipe, but not excessively high resulting in significant hydrogen embrittlement concerns.
Presence of sulfides	• Sulfides contribute to hydrogen damage in two ways: • First, the hydrogen sulfide reacts with the iron to form atomic hydrogen – increasing concentrations of the element. • Second, the presence of sulfides impedes the recombination of hydrogen atoms into hydrogen molecules so that the hydrogen atoms are left on the surface of the steel and have more time to be absorbed into the metal.
Welding process	• Hydrogen may be introduced into high-carbon steels and low-alloy steels (and many other alloys) by the welding process. Specifically electrode coatings and flux can be sources of hydrogen.

hydrogen embrittlement: See *hydrogen damage*.

hydrogen flux monitoring: See *hydrogen monitoring*.

hydrogen-induced cracking: See *hydrogen damage*.

hydrogen foil: See *hydrogen monitoring*.

hydrogen monitoring: A method that establishes whether active corrosion is present based on measuring hydrogen levels. On pipelines, hydrogen monitoring is usually called "hydrogen flux monitoring" because it monitors internal corrosion by detecting the diffusion of hydrogen through the pipe wall.

One of the main methods of hydrogen monitoring is the measurement of pressure change on the external pipe surface. More specifically, atomic hydrogen is produced by the cathodic half-cell reaction at the internal surface of the pipeline. Some of that hydrogen will diffuse through the pipe wall and is trapped as molecular hydrogen between the external surface of the pipeline and a stainless steel foil attached to the external surface. See Figure 55. Pressure between the foil and pipe wall is monitored for increases in pressure. An increase in pressure may indicate active corrosion on the internal surface of the pipeline.

hydrogen stress cracking: See *hydrogen damage.*

hydrogen sulfide (H_2S): A colorless gas that is poisonous to humans and animals. Also known as hydrosulfuric acid, sewer gas, and stink damp, it is recognizable by its rotten egg smell at very low concentrations (0.01 – 0.3 parts per million). Table 48 summarizes the key effects of sour gas on humans.

There are also implications for pipeline design from a pipeline integrity (i.e., internal corrosion and cracking) point of view. As such, both safety and materials considerations are significantly greater for a pipeline carrying sour product. Specifically, in environments with hydrogen sulfide (H_2S) present, there is heightened potential for corrosion, sulfide stress cracking and hydrogen damage. Pipelines carrying high concentrations of H_2S often special design and maintenance requirements. CSA (5.4.1) defines the threshold for sour service (associated with more stringent material requirements) for gas pipeline systems, as service in which the hydrogen sulfide partial pressure exceeds 0.35 kPa. However, the presence of other substances (such as carbon dioxide) can influence this guidance. Both CSA Z662 and Standard MR0175: Sulfide Stress Cracking Resistant Metallic Materials for Oilfield Equipment apply provide additional guidance.

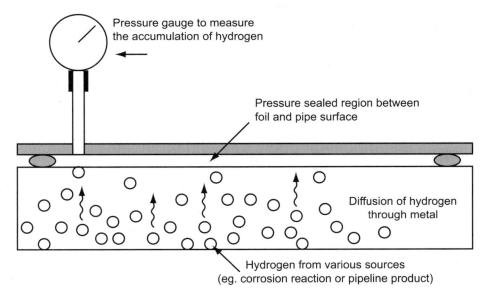

Figure 55. Principle of hydrogen flux monitoring.

Iron sulfides are often formed from corrosion reactions, and can be important in corrosion control, especially at lower temperatures and low H_2S partial pressures, where a protective film often forms. However, for this protective film to form, the absence of oxygen and chloride salts is required. Other corrosion products that may form include black or blue-black iron sulfides, pyrite, greigite, mackinwaite, kansite, iron oxide (Fe_3O_4), magnetite, sulfur (S), and sulfur dioxide (SO_2).

hydrogen-induced cracking (HIC): See *hydrogen damage*.

hydrostatic test: A process where the pipeline is filled with a liquid (e.g., water), sealed, and subjected to a pressure higher than the normal operating pressure using dedicated pumps or compressors. See *pressure test* for additional detail.

hydrotest: See *pressure test*.

Table 48. Effect of H_2S on humans [adapted from (11)]

Parts per million (approx. ranges)	Effect on humans
0.01-0.3	Odor is detectable
1-10	Moderate to strong odor. People may experience nausea, tearing of the eyes, headaches, and loss of sleep following prolonged exposure. The effects appear to be reversible and not serious for the general population, although more susceptible individuals may respond more severely.
10-150	Increasing degree of irritation to eyes and lungs; affects sense of smell and as a result there is no perceptible odor.
150-750	Severe health effects that may lead to death become more likely as concentration and exposure time increase.
greater than 750	Death may occur in minutes or less.

I

ice scour: Refers to the long thin gouging of the seabed that can occur from the movement of ice. The phenomenon can result in significant deformation of the seabed and may present a significant hazard to subsea pipelines if not buried to sufficient depth.

ID: Internal diameter of a pipeline.

in-line inspection: See *in-line inspection.*

in-line inspection log: See *in-line inspection.*

incomplete penetration: See *lack of penetration* and *weld defects.*

industry code: See *code.*

immediate repair condition: An anomaly that meets specific criteria and requires immediate action (i.e., pressure reduction and/or repair of the pipeline). The criteria for immediate repair condition appear in CFR 195.452 for liquid lines and §192.703(c) for gas lines.

IMP: See *integrity management plan.*

impact damage: Refers to harm caused to either the pipeline itself or the pipeline coating as a result of a large/hard object making forceful contact with the pipe. The term is used interchangeably with the term *mechanical damage* and is largely used in the context of construction practices or *3rd party damage.*

impact radius: See *potential impact radius.*

imperfection: An *anomaly* in the pipe that will not result in a pipeline failure at loads anticipated during the life of the pipeline.

impingement corrosion: A form of erosion-driven corrosion where a high-velocity flowing fluid impacts a solid surface in a very localized area.

impressed-current cathodic protection: One of the two principal types of cathodic protection where an external power source is used to polarize the pipeline and to deliver current into the soil or water. See *cathodic protection*.

incident: The term incident refers to a broad range of upset conditions relative to normal operations. The term has various applications, such as human safety, environmental impact, and uncontrolled release of product, or operation beyond design limits. It should be noted that this term usually has a specific regulatory definition that varies by jurisdiction.

incident management system: The systems, processes, and procedures that form the basis for a coordinated and timely response to an *incident* to minimize the impact to the public, the environment, and company operations. The outcomes obtained from these processes results in tactical approaches in dealing with incidents that are documented in an *emergency management plan*.

inclusion: **1.** A foreign (usually) non-metallic material in the pipe wall – introduced during the fabrication process. Inclusions are usually oxides, sulfides, or silicates, but they may be of any substance that is a foreign material. **2.** An inclusion is formed when foreign material (debris, metal slivers, etc.) becomes incorporated into pipeline coating.

indication: An unconfirmed possible deviation from sound pipe material or a weld. An indication may be generated by a non-destructive inspection, such as an in-line inspection. This term is often used interchangeably with the term *anomaly*.

indirect inspection: **1.** An inspection method that is not *direct* and therefore relies on inference to determine the state of the pipeline. For example, a visual inspection of the pipeline would be considered a direct inspection whereas taking cathodic protection readings to identify corrosion would be considered indirect inspection. **2.** The second step in the *direct assessment methodology,* which guides the operator in choosing direct assessment locations. See *direct assessment methodology*.

individual risk: This term is most often used in the context of quantitative risk and reliability based methods for establishing maintenance programs. CSA Z662 (Annex O) defines it as "The annual probability of fatality due to a pipeline incident for an individual situated at a particular location." This is in contrast to *societal risk*.

induction coil: A type of sensor that measures the time rate of change in the magnetic flux density – most commonly used in *standard resolution MFL tools*. Induction coils do not require power to operate, but have a minimum inspection speed requirement.

inductive resistance corrosion monitoring: Closely related to *electrical resistance corrosion monitoring*, but it is much more sensitive. The resistance of the material is measured by inducing a current in the metal by an AC (alternating current) coil.

inertial tool: This term refers to a *pig* that is fitted with an inertial navigation system. Specifically, the tool uses an onboard computer and a combination of motions sensors (e.g., gyroscopes and accelerometers) to establish a very accurate three-dimensional profile of the pipeline. Inertial tools are often combined with caliper arms to detect and measure dents and other forms of mechanical damage.

INGAA: See *Interstate Natural Gas Association of America*.

inhibitor: A chemical substance or combination of substances that will reduce corrosion rates when present in small concentrations. Corrosion inhibitors are an important method of controlling internal corrosion in both gas and liquid pipelines where a significant amount of water is present. Inhibitors can be classified various ways; however, the most common classification is based on how they function, as shown in Table 49. The manner in which inhibitors are applied is highly dependent on the class of inhibitor chosen and the operating conditions of the pipeline. However, there are two main approaches: application in a *batch* or through continuous injection.

inhibitor efficiency: A measure of effectiveness of a corrosion inhibitor in reducing corrosion rate. The efficiency, E, in percent, is calculated:

Table 49. Most common inhibitor types for pipeline applications and the associated mechanism. (18)

Inhibitor type	Mechanism
Passivating	• Reduce corrosion through anodic shift moving the metal into the passivation range. • Two main classes: oxidizing (do not require presence of oxygen) and non-oxidizing (require the presence of oxygen) to passivate steels. • One of the most effective and inexpensive classes of inhibitors.
Cathodic	• Cathodic inhibitors impede the cathodic reaction either by one of three mechanisms: ◊ Cathodic poisoning (slowing down the cathodic reaction). ◊ Cathodic precipitate (creating a physical barrier to the cathodic reaction). ◊ Oxygen scavenging (reducing availability of oxygen for corrosion reactions).
Organic	• Organic inhibitors work by forming a hydrophobic film over the surface of the metal to be protected.
Precipitation	• Precipitation inhibitors work by encouraging a chemical reaction in the pipeline product that results in a compound that is not soluble – the compound then creates a protective layer on the metal through deposition.
Volatile	• Also termed "vapor phase" inhibitors; they are transported to the areas susceptible to corrosion (in a closed system) through vaporization at the source. • Not very common in pipeline systems.

$$E = 100\frac{r_U - r_I}{r_U}$$

where r_u = uninhibited corrosion rate
r_i = inhibited corrosion rate

in-line inspection: The process of determining the condition of a line, with respect to a specific defect type (e.g., cracking, corrosion, deformation, etc.), based on the results of using an *in-line inspection tool*. See *pig*.

in-line inspection report: The results of an in-line inspection provided by the vendor identifying the detectable anomalies in a pipeline section. The specifics of reporting times and size/nature of defects identified vary significantly based on the specific contractual agreement, the in-line inspection technology used, the density of anomalies in the pipeline, the pipeline length, and diameter. For an extract of typical reporting for corrosion inspections, see *pipe tally*.

in-line inspection tool: One of two main classes of *pigs* that provide information on the condition of the line as well as the extent and location of any problems. The exact nature of the information collected is dependent on the technology selected; however, generally, the tool itself is simply the device that gathers the data – the data must then be analyzed to determine and report on the condition of the line. See *pig*.

in-service: Used interchangeably with the terms "on-stream" or "on-line" referring to the state of a pipeline system as being fully operational. Specifically, the term is relevant when one refers to a (maintenance) activity or modification of the pipeline that can be carried out while the pipeline remains in full operation – that is, the pipeline remains "on stream."

inspection interval: The time interval between which a pipeline segment is inspected for anomalies (i.e., corrosion or cracking) using various monitoring methods; however, the term is most commonly used with reference to the frequency of inspection using a smart *pig*. While various approaches can be taken in establishing inspection intervals (e.g., based on deterministic calculations vs. probabilistic analysis), the key factors to consider (and their potential implications) remain common to all approaches. Table 50 shows the key determinants and associated considerations.

inspector: See *owner-inspector*.

instant "off" potential: Refers a reading of the electrical potential done immediately following the drop in potential after the impressed current cathodic protection system is turned off. The instant "off" potential is measured to establish the effective protection of a pipeline using either the 100 mV or the -850 mV criteria.

Table 50. Key considerations in establishing an inspection interval for smart pigs

Factor	Consideration
Cost	• The cost of a repeat in-line inspection can often be lower than the cost of repairing all indicated anomalies – especially in pipelines with a high defect density where the operator is faced with an extensive repair program.
Corrosion/crack growth rate	• The expected/estimated rate of corrosion or crack growth is a key factor in determining the frequency of re-inspection because the higher the corrosion rate, the more frequently one can expect anomalies to grow to failure pressure.
Consequence	• If the consequence of failure is high, one would expect to inspect more frequently to ensure that there is greater certainty associated with the condition of the pipeline.
Corrosion/crack density	• More densely flawed lines are likely to be inspected more frequently for two reasons. First, defect sizing can be more difficult in the presence of complex defect geometries reducing measurement certainty; second, a greater number of excavations are likely to be required, which may result in a repeat inspection as being a more cost effective option to increase certainty in the condition of the line.
Time based limitations of defect growth analysis	• Even if no anomalies are expected to fail, due to the uncertainty associated with tool defect measurement and sizing, one would expect that the maximum time between inspections is 7 to 10 years – depending on the specific context.

integrity management plan (IMP): A document that itemizes the underlying philosophies and corresponding activities that are needed to ensure safe long-term operation of a pipeline system or network. That is, the documentation of a systematic approach to prevention, detection, and mitigation of pipeline integrity related concerns on a five-to-ten-year time horizon. It should be noted that the scope is not limited to engineering/field activities, but is inclusive of data management, training, and change management requirements.

integrity management program (IMP): The culmination of the systems and processes used to determine the key activities required to maintain safe long-term operation of a pipeline system. It is the set of specific activities that occur on an annual basis to prevent, detect, and mitigate pipeline integrity related concerns.

interaction rule: A specification that establishes spacing criteria between anomalies or defects that must be considered together for the purposes of calculating failure pressure. That is, if the indications or defects are proximate to one another within the criteria, the anomaly or defect is treated as a single unit for the purposes of engineering analysis. Several approaches exist including *3t × 3t* and minimum length × minimum width.

interference: See *stray current*.

i

intergranular corrosion: A form of corrosion that occurs preferentially at or along the grain boundaries of a metal. The term is also used interchangeably with intercrystalline corrosion.

intergranular stress corrosion cracking: This term is often used interchangeably with the term *classical SCC*. See s*tress corrosion cracking*.

interlinking cracks: A scenario where individual cracks have grown in a manner such that they have begun to join together, or coalesce. The key implication is that each crack cannot be assessed in isolation, because failure in the vicinity would be affected by a number of cracks that interact in some manner.

internal corrosion: Corrosion encountered on the inner surface of the pipe (exposed to the product flow). See *corrosion*.

internal diameter: The actual diameter of the pipe available for product flow. See diagram for *outer diameter*.

internal liner: See *liner*.

internal probe: A measuring device or instrument inserted in the pipeline to measure a flow property, such as pressure or temperature. Internal probes can be of concern during in-line inspection programs because they may protrude into the pipe presenting an obstacle to the safe passage of a tool. As such, the presence of probes should be identified as part of a pre-run questionnaire and removed for the actual inspection to avoid probe and tool damage.

International Standards Organization (ISO): ISO is a network of national standards institutes from 163 countries – the network is coordinated through a central Secretariat in Geneva Switzerland. The organization provides a mechanism to develop consensus based standards (and standards type documents) for technical topics across a broad range of industries and is the largest organization of its kind globally: the network encompasses 3,274 individual technical bodies (technical committees, working groups and ad hoc study groups). (190)

Interstate Natural Gas Association of America (INGAA): A trade organization that advocates regulatory and legislative positions of importance to the natural gas pipeline industry in North America. Member companies are responsible for transporting over 95% of the natural gas in North America on a network that is about 200,000 miles long. Additional information is available at www.ingaa.org.

intragranular corrosion / cracking: A form of corrosion that does not preferentially corrode at, or along, the *grain* boundaries of a metal – rather the growth pattern is essentially unchanged by the presence of *grain* boundaries.

ion: An atom or group of atoms (i.e., molecule) that has gained or lost one or more electrons and thus carries an electric charge. Positive ions, or cations, are deficient in outer electrons. Negative ions, or anions, have an excess of outer electrons.

iron sulfide: A corrosion product found in pipelines where a significant amount of hydrogen sulfide is present. Iron sulfide can be pyrophoric and spontaneously ignite when exposed to air. It is therefore is a hazard during pigging operations if the pig emerges from the pipeline with iron sulfide.

ISO: See *International Standards Organization*.

isolation valve: Usually, part of a larger grouping of valves and used to segregate a portion of the pipeline from the remainder of the system. Isolation valves are installed to allow the segregation of a (relatively) small section of pipe to facilitate construction or maintenance activities that cannot be carried out at normal operating conditions (i.e., a pressure reduction or evacuation of the pipe is required). Isolation valves also play a key role in limiting product release and consequential damage resulting from a pipeline leak or rupture by limiting the produce flowing to the affected area.

J

Japanese Industrial Standard (JIS): Japan has unique pipe specification standards defined by the Japanese Industrial Standard, which specifies the standards used for industrial activity in the country. Associated documentation is published by the Japanese Standards Association.

jeep: An instrument used to detect coating defects (*holidays*). See *holiday detection*.

JIS: See *Japanese Industrial Standard.*

joint: A length of pipe formed from a single piece of steel at the pipe mill. Pipelines are built by welding many joints of pipe together. For pipe transported by ship, joints are typically 40 ft (12 m) long; domestically manufactured pipe in North America can be longer than 72 feet (22 m). A section of pipe that is smaller than the typical length of a join is called a *pup*.

jumper: A short flowline connecting a subsea well back to its manifold.

K

kicker line: A term that describers the piping and valves that connects the pressurizing pipeline to the launcher or receiver. Specifically, the kicker lines facilitate the movement of the pig between the barrels and the main pipeline segment that is to be inspected. See *pig traps* for detail regarding the connection points of the kicker lines.

kilometer post (kp): Linear distance of a point along the pipeline. Typically, the distance is measured from a fixed reference such as a compressor/pump station, valve, or pig launcher. Used interchangeably with the term *chainage*.

knife-line attack: The intergranular corrosion of an alloy (usually stabilized stainless steel) along a line adjoining or in contact with a weld after heating into the sensitization temperature range.

kp: See *kilometer post*.

L

lack of cover: A situation where an insufficient amount of soil or other material covers a pipeline. In extreme cases the pipe may actually be exposed, resulting in a higher risk for third-party damage and excessively high strains, coating degradation, heat and/or UV damage. A lack of cover can be caused by several situations including, but not limited to, erosion, buoyancy problems, and ground movement.

lack of fusion: A welding defect where the weld material fails to fuse with the pipe wall or previous bead. In manufacturing, lack of fusion can occur on *electrical-resistance welded* (ERW) pipe; the lackof fusion anomalies are notch-like faults that run down the center of an ERW seam weld.

lack of penetration: A situation where the welding process has resulted in incomplete deposition of weld metal at the bottom of the weld joint. Specifically, a gap remains between the two components being joined that should have been filled with weld metal. Depending on the severity of the defect, code requirements will indicate whether the weld requires repair. Figure 56 illustrates a lack-of-penetration defect in a weld.

The edges of the pieces have not been welded together, usually at the bottom of single V-groove welds.

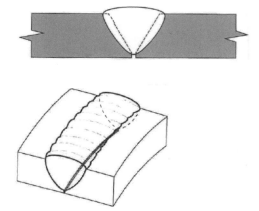

Figure 56. Lack of penetration in weld. (197)

laminar flow: The flow of fluid (i.e., liquid or gas) in which the flow paths are in smooth, parallel lines, with practically no mixing and no turbulence. See *turbulent flow* for a visual representation of both turbulent and laminar flow. For a discussion regarding multi-phase flows, see *flow regime*.

lamination: A manufacturing feature. A lamination is a planar discontinuity in the metal usually parallel-to-nearly-parallel to the surface of the pipe wall. Laminations may be caused by the non-metallic inclusions, which are rolled into the metal during manufacturing. Surface-breaking laminations are laminations that intersect the surface of the metal.

land agent: Individuals who acquire and manage land rights for pipeline-related activities. These individuals interface closely with landowners along the right-of-way as well as pipeline personnel (e.g., operators, contractors, surveyors, environmentalists, and maintenance crews).

landowner: An individual who resides upon, or owns, land adjacent to, or in the vicinity of, the pipeline ROW or related facilities.

lap: A manufacturing anomaly. It is a lamination not fully parallel to the surface of the pipe, or where a metallic flap is rolled into the surface of the pipe wall.

laser mapping: The measurement and documentation of external corrosion depth on a pipeline by using laser technology. Specifically the tool records information from the reflected laser light to build a two-dimensional profile of the pipeline surface; this information is then integrated with location information to produce a three-dimensional profile of the pipeline.

launcher: See *pig trap*.

leak: A pipeline leak is a condition where a quantity of the pipeline product is released into the environment in an uncontrolled manner through a (relatively) small area of pipe wall that has been compromised, or through valves, flanges, fittings, etc.

leak detection: This refers to tools and techniques employed by pipeline operators to identify and locate the unplanned release of pipeline product. The most common methods used include *mass balance, aerial survey,* ground patrols, and *remote sensing*.

life cycle costing: This is the calculation of cost of construction, inspection, maintenance, ownership, and abandonment of an asset. The cost calculation is governed by the Generally Accepted Accounting Practices (GAAP) and is calculated as a net present cost Life-cycle costing is often used to compare alternative projects being contemplated.

life cycle management: A management principle aimed at maximizing the return on investment of an asset over its entire life (i.e., design through to decommissioning). The emphasis of life cycle management is the long-term value rather than short-term returns.

lift-off: See *stand-off.*

limit states design: Limit states design is a structural engineering philosophy that augments conventional load limit designs (i.e., where the maximum allowable pipe stress is the limiting factor) with consideration for a limit for every potential mode of failure such that the critical factor driving design considerations may be something other than hoop stress (e.g., strain limits may be relevant where the maximum allowable strain is the critical factor). The development of this approach has largely been driven by the difficulty of cost effectively meeting conventional design requirements in environments where significant longitudinal strain (typically >0.5%) of the pipeline is anticipated (e.g., polar/permafrost regions, laying offshore pipe, and unstable slopes). The approach allows selective extension of the stress-based design options to take advantage of the fact that steel can deform plastically yet remain structurally stable (i.e., *strain based design*).

line location: The process of surveying the exact location of a buried pipeline to facilitate safe excavation (and other maintenance activities) in the vicinity of the pipeline right-of-way.

liner: A hollow tube of (usually) plastic or polyethylene that is installed in the interior of an existing pipeline. The main use of liners – from a pipeline integrity perspective -- is to provide a barrier to prevent corrosive process fluids from contacting the steel carrier pipe, which allows reduced mitigation requirements versus corrosion mitigation and monitoring programs using carbon steel line pipe. Liners can also be used to improve flow characteristics (to reduce energy consumption). Recently, metal-forming liners have been used to restore pressure capacity of damaged pipelines.

liquefied natural gas (LNG): Natural gas that has been cooled (to a temperature ~-250°F) to a liquid state. The liquefaction of natural gas is often considered to ease transport costs (via ships and/or vehicles) where pipelines are not a practical option (i.e., across large bodies of water). LNG facilities do represent a significant investment and has only become more prevalent in recent times where natural gas supply and markets are separated by large (oceanic) distances.

liquefied petroleum gas (LPG): Refers to a range of hydrocarbon gas compositions in which the primary constituents are propane or butane.

liquid couplant: See *couplant.*

liquid penetrant inspection (LPI): A non-destructive testing technique used for surface-breaking anomalies such as cracks. This method is used primarily for non-ferromagnetic material, because *magnetic-particle inspection* (MPI) is effective on ferromagnetic materials such as pipeline steels and iron. The main steps involved in LPI are shown in Figure 57. See *penetrant* for an overview of the types of liquids that are used in this technique.

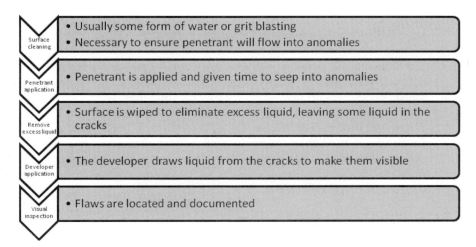

Figure 57. Key steps in liquid penetrant inspection. (1)

LNG: See *liquefied natural gas.*

load and resistance factor design: One of the methods of pipeline design that lies on the spectrum between a fully deterministic approach and a reliability-based approach. See Figure 58. Thus, the LRFD design reflects greater knowledge regarding both the material properties and expected pipeline loads. It is essentially a largely deterministic application of *limit states design.*

local cell: An electrochemical cell that results from the inhomogeneity on a pipe surface immersed in an electrolyte. The inhomogeneities may be of a physical or chemical nature in either the metal or its environment. For example, the existence of a bacterial colony can produce a local cell.

localized corrosion: Refers to degradation small discrete areas as opposed to general corrosion. Examples of localized corrosion include pitting, crevice corrosion, and stress corrosion cracking.

lognormal distribution: A probability distribution defined on the interval $(0, \infty)$. The distribution is derived from a normal distribution in that with a lognormal distribution the natural log of x, $\ln x$, has a normal distribution. Mathematically, the PDF of the lognormal distribution, $LogNorm(x)$, is defined as

$$LogNorm(x) = \begin{cases} \dfrac{1}{x\sqrt{2\pi\sigma^2}} e^{-\frac{(\ln x - \mu)^2}{2\sigma^2}} & x > 0 \\ 0, \; x \le 0 \end{cases}$$

where μ = location parameter
                 σ = shape parameters of the distribution

Table 51 shows selected properties of the lognormal distribution. Figure 59 illustrates several lognormal distributions for various parameters.

The lognormal distribution is often used to model the estimates of ratios or probabilities and parameters which are nonnegative such as dimensional parameters (i.e., wall thickness).

Lognormal distributions have a few applications in the analysis of in-line-inspection (in-line inspection) data. An estimate of the proportion of in-line inspection depths that are within tolerance can be modeled as a lognormal distribution. Lognormal distributions have also been used as the probability distribution of the true depth given a measured corrosion depth.

long seam: See *long-seam weld.*

longitudinal wave: See *compressional* wave and *ultrasonic testing* for a more detailed discussion of the topic.

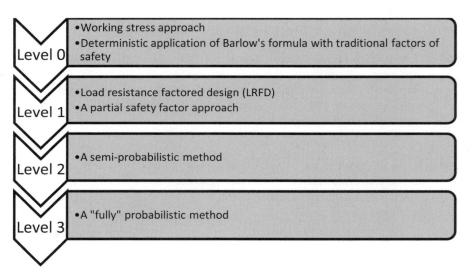

Level 0
- Working stress approach
- Deterministic application of Barlow's formula with traditional factors of safety

Level 1
- Load resistance factored design (LRFD)
- A partial safety factor approach

Level 2
- A semi-probabilistic method

Level 3
- A "fully" probabilistic method

Figure 58. Methods of pipeline design. (9)

Table 51. Lognormal distribution properties

Parameters	μ and σ
Domain	$0 \leq x \leq \infty$
Mean	$e^{\left(\mu + \frac{1}{2\sigma^2}\right)}$
Variance	$e^{2\mu + \sigma^2}\left(e^{\sigma^2} - 1\right)$
Mode	$e^{\left(\mu - \frac{1}{\sigma^2}\right)}$

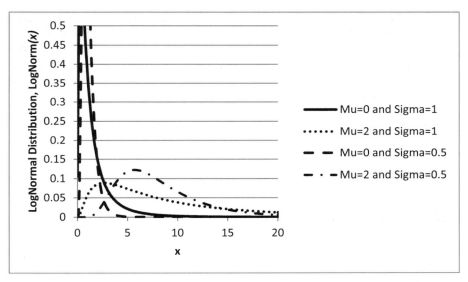

Figure 59. Lognormal distribution for various values of μ and σ. The special case of α = 1 is the exponential distribution.

long-seam weld: Refers to the longitudinal join that is required during most pipe fabrication processes. Most pipe fabrication processes involve the use of steel plate that is rolled into a hollow cylinder and then joined (welded) along the long axis of the cylinder – creating a long-seam weld.

loop line: A colloquial term that indicates an additional, or new, length of pipe that parallels an existing pipeline.

loss of adhesion: Refers to a type of coating failure where large areas of the coating become disbonded from the pipe surface, increasing the likelihood of corrosion and cracking on the pipe surface.

love wave: A specific form of transverse ultrasonic wave that travels along the surface of the material being tested (in this case to the depth of up to several wave lengths). In the context of pipeline applications, love waves have been used primarily for the detection and sizing of cracks. See *ultrasonic testing*.

low pH SCC: See *transgranular SCC*.

low-pressure water jetting: The term is used interchangeably with low-pressure water cleaning. See *water jetting*.

low-resolution in-line inspection: Refers to the first generation of in-line-inspection tools. These tools have a limited number of sensors, which reduces their detection and sizing abilities. Usually, reports from low-resolution inspections would only grade an anomaly in broad depth ranges. One such grading is shown in Table 52. Low-resolution tools have been supplanted by the introduction of *high-resolution in-line inspection* in recent years.

LPG: See *liquefied petroleum gas.*

LPI: See *liquid penetrant inspection.*

LPR: See *linear polarization resistance.*

LRFD: See *load and resistance factor design*

Table 52. Tool depth grading for Standard-resolution MFL tools

Grade	Depth Range
Light	<25% NWT
Moderate	25-40% NWT
Severe	>40% NWT

M

magnesium anode: An anode primarily composed of magnesium, best suited to *galvanic cathodic protection* systems where low current requirements are acceptable.

magnetic circuit: See also *magnetic flux leakage.*

magnetic-cleaning pig: Usually, a single-module utility pig mounted with magnets to facilitate removal of magnetic debris from the pipeline (e.g., welding rods). A magnetic cleaning pig is the least aggressive cleaning pig. The magnets are mounted near the center of the pig and are used to collect and hold magnetic debris. Magnetic-cleaning pigs are often run as part of cleaning program to prepare for an in-line inspection run or prior to the application of an inhibitor batch. See *pig.*

magnetic flux density: See *magnetic flux leakage.*

magnetic flux leakage (MFL): A technique that uses magnetic fields to detect corrosion and other faults in pipelines; *pigs* capable of applying this technique are now commonly used to identify and size the metal loss on that transport hydrocarbon products. The technology uses a multi-segment pigging tool with a magnetizer and sensors in one segment, and data storage components, power supply, and other sensors in additional segments.

The magnetizer segment encompasses the magnetic circuit (see Figure 60) of the tool and contains powerful permanent magnets that magnetize the pipe wall, usually through steel brushes, as the pig travels through the pipe. The magnetic circuit saturates the pipe wall with magnetic flux so that any discontinuity in the metal will cause magnetic flux to exit the metal. This disruption or "leak" of the magnetic flow is detected by sensors located around the circumference of the tool, and the data can be interpreted to establish the dimensions of the metal loss. Secondary sensors make it possible to discriminate between internal and external metal loss (not shown in figure).

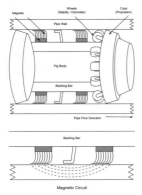

Figure 60. Simplified schematic of magnetic flux leakage pig.

magnetic particle inspection (MPI): A non-destructive technique to locate cracks and other anomalies that are open to the surface being inspected. The method is restricted to the inspection of ferromagnetic material, such as steel and iron, and consists of several steps, as shown in Figure 61.

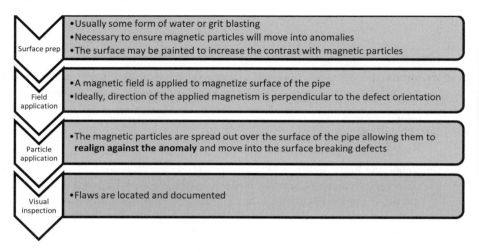

Surface prep	• Usually some form of water or grit blasting • Necessary to ensure magnetic particles will move into anomalies • The surface may be painted to increase the contrast with magnetic particles
Field application	• A magnetic field is applied to magnetize surface of the pipe • Ideally, direction of the applied magnetism is perpendicular to the defect orientation
Particle application	• The magnetic particles are spread out over the surface of the pipe allowing them to **realign against the anomaly** and move into the surface breaking defects
Visual inspection	• Flaws are located and documented

Figure 61. Summary of magnetic particle inspection process. (1)

magnetic saturation: The state where a sufficiently large magnetic field is applied to a magnetic material, such as steel, to fully magnetize the material. Any further increase to the magnetic field will not increase the magnetization of the material. For a conventional MFL inspection, the purpose of the permanent magnet on the tool is to magnetically saturate the pipe wall. MFL technology takes advantage of fully saturating the pipe wall magnetically, because beyond this point, any change in the magnetic field is due to changes in geometry only – changes in material properties will not affect the magnetic field measurement. Changes in physical properties of the material (e.g., hardness) are detected only if the magnetization level is below saturation.

main-line valve: Any one of the valves located on the main section of a pipeline spaced at regular intervals. The purpose of these main-line valves is to provide a mechanism for isolating a section of pipe in the case of a failure or to facilitate construction and maintenance activities.

Major Industrial Accidents Council of Canada (MIACC): Established in 1987 following the Union Carbide industrial disaster in Bhopal, India, MIACC was formed to investigate if a similar accident could happen in Canada; however, during its existence, the organization provided guidance on how to assess public safety associated with the release of hazardous substances associated with industrial activity. While the committee has since been disbanded, its work is often used in the context of establishing safety thresholds when undertaking (absolute) quantified risk assessments of pipeline safety.

mandrel pig: A type of utility pig. See *pig* for a detailed discussion of this class of tools.

manufacturing defect: See *mill related anomalies* and *weld defects*.

MAOP: See *maximum allowable operating pressure.*

mapping tool: A smart pig that uses inertial sensing or other technology to collect data that can be analyzed to produce an elevation and plan view of the pipeline centerline. See *pig*.

martensite, martensitic steel: One of several microstructures formed in the making of steels (other relevant ones being ferrite, austentite, and pearlite). Martensite is the structure responsible for the hardness and strength of traditionally quenched and tempered steels.

mass balance: A methodology, primarily for liquid pipelines, based on calculating the difference between product intake and product off-take volumes to identify potential pipeline leaks. The approach is based on the principle of mass conservation (i.e., mass cannot be spontaneously created or destroyed). That is, the amount of product a pipeline system must ultimately balance with the amount of product leaving the pipeline system – therefore any volumes unaccounted for must be the result of the loss of containment somewhere on the pipeline system.

maximum allowable operating pressure (MAOP): Refers to the maximum internal pressure legally permitted during its operation. This value is usually the same as the design pressure unless the pipeline is been derated.

maximum operating pressure (MOP): The maximum pressure within a pipeline during steady-state operations. While this pressure can be less than the maximum allowable operating pressure of a pipeline due factors such as operational limitations (e.g., elevation), it cannot exceed the maximum allowable operating pressure for all but very short periods of time. The value is usually determined over a relatively short historical time frame (i.e., maximum pressure that occurs during normal operations over a period of one year).

mean: The arithmetic average of a data set – not to be confused by the mode, median or "middle" of data sample. The mathematical formula varies if the data set is described by a specific probability distribution (see various entries for probability distributions). However, for a specific sample of data, the mean can be calculated as follows:

$$\bar{x} = \frac{1}{N} \left(\sum_{i=1}^{N} x_i \right)$$

Where $\bar{x}$ = mean of the sample data set
N = number of data points in the sample
x_i = value of the specific data point

measurement threshold: This term is most often used in the context of in-line inspection tools but would equally apply to any tool that reports direct or inferred measurements. The term refers to the lower limit below which a tool is unable to detect and/or reliably take measurements. The lower limit is typically the result of a physical limitation (i.e., the size of tip of a pencil probe may not allow accurate depth measurement of a pit that has a diameter narrower than the probe itself). In the case of inferred measurements, the lower limit may be the result of noise in the signal. For example, with high-resolution MFL tools, measurements below a depth of 10% are treated as being less reliable due to the level of noise in the signal.

mechanical damage: This term refers to anomalies and defects on a pipeline, such as dents, gouges or localized areas of cold work, which are caused by external force acting on the pipeline. Mechanical damage is often associated with coating damage, significant pipeline deformation, and high residual stresses.

Mechanical damage on pipelines is most commonly the result of construction or third-party activity once the pipeline has been established. Mechanical damage poses a challenge to pipeline operators because of the difficulty in assessing its severity. Part of the difficulty stems from the fact that mechanical damage can often alter the metallurgical structure of the steel. That is, localized hardness can leave the material more susceptible to cracking.

mercaptan: See *odorant*.

metal loss (ML): An anomaly identified by an in-line inspection (in-line inspection) where metal has been removed from the pipe wall or where the wall thickness has been reduced. Metal loss is most often applied to corrosion anomalies; gouges are also metal-loss anomalies. Mill anomalies, such as inclusions where metal is missing, are often referred to as metal loss, despite that no metal was actually lost. See *defect classification*.

MFL: See *magnetic flux leakage*.

MIACC: See *Major Industrial Accidents Council of Canada*.

microbiologically influenced corrosion (MIC): Corrosion that is influenced by the presence and activity of various microorganisms, such as *bacteria*, fungi, algae and protozoa. The mechanism through which bacterial colonies are established starts where small free-floating organisms settle, and attach, themselves onto a metal surface. Over time, the number of organisms living on the surface is sufficiently large to create a biofilm (i.e., continuous layer of organisms) on the surface of the metal. Once the initial biofilm has been established, it creates an environment where organisms that are a specific concern in MIC, such as sulfate-reducing bacteria, can survive and affect corrosion. MIC influences corrosion in pipeline primarily by four mechanisms: acid production, sulfate reduction, differential aeration cells, and selective dissolution.

On pipelines, microorganisms can affect internal and external corrosion. Sulfate-reducing bacteria (SRB) are the most significant problem. They exist in most soils and can influence external corrosion. Water-saturated clay soils, near neutral pH, decaying organic material with a source of SRB are particular problem areas. The cathodic protection systems can create an anaerobic environment near the pipeline where SRB can flourish.

Detecting MIC can involve the assessment of several indicators including the presence of water, nutrients, oxygen, etc.; the morphology of the corrosion; and the nature of the corrosion products. However, final verification of the presence of microbes is typically by direct inspection by microscope or by growth assays.

The primary form of mitigation for MIC found on the internal pipe surface is pipe cleaning (disrupts the formation of a biofilm) and use of biocides (kills the microbes). Biocides (i.e., inhibitors) can be injected continuously or run in a batch. The primary form of mitigation for MIC found on the external pipe surface is coatings and cathodic protection.

midwall feature: A feature, introduced through the fabrication process that is not surface breaking. The most common examples include *laminations* and *inclusions*.

MIG welding: Refers to gas metal arc welding (GMAW). See *welding*.

mile post: See *kilometer post*.

Mill-related anomalies: Imperfections in the pipe that are introduced through the fabrication process. These anomalies may or may not be injurious. A brief summary of typical (non-weld) mill-related anomalies (see *weld defects* for weld related mill anomalies) appears in Table 53.

mill scale: Refers to the heavy oxide layer (i.e., corrosion product) that is formed during the hot fabrication or heat treatment of metals. As part of good surface preparation, the scale must be removed prior to the application of any protective coating.

Table 53. Common mill-related anomalies (8, 10, 195)

Defect	Description
Blisters	A gap or void in the metal that is fully enclosed by a layer of steel. See *blister*.
Center line segregations	A thin layer of non-metallic impurity in the pipe wall – this is often a result of this material collecting in the center of an ingot where the material solidified last.
hard spot	A localized area in the pipe material with higher strength (lower ductility) resulting from rapid cooling/quenching (this can be the result of poor quality control in the fabrication process.
Hook cracks	A lamination in the plate used to form pipe that is deformed during the fabrication process to form a hook crack in the long seam weld area. See hook crack.
Inclusion	A generic term describing a situation where impurities have been incorporated into the pipe metal. See *inclusion*.
indentations	The deformation of the pipe surface (i.e., impression left) from foreign material that is trapped between the rolling surface and the plate. These are usually non-injurious.
laminations	A two dimensional discontinuity parallel to the surface of the pipe occurs when inclusions or concentrations of non-metallic contaminants are rolled out (i.e., flattened) during the fabrication process.
lap	A lamination that might not be fully parallel to the pipe surface effectively leaving a "flap" of material in place.
sliver	An area where a lap has detached resulting in an area Detached lap – leaves shallow area of thinner wall.

misalignment: A weld where the ends of the two pipe joints have been misaligned in some way – potentially compromising the integrity of the weld. Figure 62 shows a misalignment at a weld.

misplaced cap: A weld where the final weld pass fails to cover the previous bead (potentially resulting in excess cap height in another area of the weld). Figure 63 shows a misplaced cap.

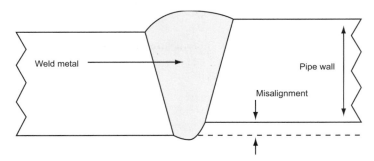

Figure 62. Misalignment at a weld.

Table 54. Major types of risk mitigation

Mitigation type	Description
Threat mitigation	Activities that mitigate a particular threat are focused on reducing the probability of an adverse event occurring. For instance, an attempt at reducing corrosion rates through cathodic protection will result in a lower likelihood of a pipeline leak – thus, cathodic protection is a type of threat mitigation.
Consequence mitigation	Activities that mitigate a particular consequence focused on reducing the damage and loss resulting from an adverse event. For instance, the implementation of an emergency management system will likely reduce the potential for injury and environmental impact from a pipeline failure, regardless of the cause of the failure – thus, the implementation of an emergency management system is a type of consequence mitigation.

missed calls: Refers to actual anomalies that exist in a pipeline but are not reported in the results from an in-line inspection run. The existence of these anomalies is addressed through the *probability of detection* function, which can be incorporated into the analysis based on in-line inspection data.

mitigation: Activities to reduce the risk exposure of a particular pipeline. Mitigation activities are broad ranging and are specific to the context (i.e., the state and conditions of the pipeline under consideration). Table 54 shows two main types of mitigation.

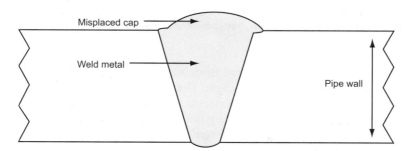

Figure 63. Misplaced cap on a weld.

mitre bend: A pipe bend that has been created when the two ends of the pipe have been cut at straight angles and then welded together; that is, the change in direction occurs abruptly at a single point. Mitre bends are typically used for small changes in the pipeline's direction and can be a concern from a pipeline integrity perspective when in-line inspection is being considered. Specifically, the abrupt change in direction at a mitre bend can result in tool damage.

modified B31G: See *B31G*.

modulus of elasticity: A property of a material that defines the relationship between an applied stress and the resulting strain (in the elastic range). Depending on the nature of the stress applied (i.e., tensile, compressive or shear), the corresponding modulus of elasticity is referred to as the *Young's Modulus*, the *bulk modulus*, or the *shear modulus*.

monitored anomaly: Refers to US regulatory requirements. A monitored anomaly is one that, once identified, does not need to be remediated; however, the operator is required to record conditions and revisit the situation on a regular basis to identify any change that could make remediation necessary.

monitoring: 1. A broad term that refers to the ongoing activity, or set of activities, associated with observing a particular parameter of the pipeline operations in support of maintaining pipeline integrity. The most common examples of pipeline monitoring activities include aerial leak surveys and routine cathodic protection surveys. **2.** Also refers to the activities needed to support the ongoing observation(s) associated with a *monitored anomaly*.

Monte Carlo simulation: A mathematical computational technique where extensive random sampling of input variables (defined by mathematical *probability distributions)* is used to develop *probability distributions* of the relevant output parameters based on a simple(r) deterministic equation. While the technique is calculation intensive (often requiring millions of iterations), it is typically used where a deterministic (equation based) solution is not available or is extremely difficult to solve. The most common application in the context of pipeline integrity is in the calculation of pipeline reliability and quantitative risk analysis. See *probability distribution*.

monumented line: A pipeline is deemed to be monumented if it has permanent survey control points established along its entire length. The permanent control points are typically bronze disks, about 10 cm in diameter, set in rock, in concrete structures, or on top of steel poles that are driven deep into the ground. These markers are then used as a basis for control of all future horizontal and vertical survey measurements and calculations associated with locating the pipeline and associated features.

MOP: See *maximum operating pressure*.

MPI: See *magnetic particle inspection*.

multi-diameter pipeline: A term used with reference to a portion of a pipeline system that is composed of more than one diameter of pipe between a launcher and receiver section. Thus, any in-line inspection, cleaning or mapping of a section requires the use of multi-diameter tools (i.e., pigs that are capable of adapting to more than one diameter of pipe). It should be noted that tools, especially smart tools, are typically only capable of handling a single increment of change in pipe diameter. Further, a multi-diameter configuration is not the norm for most inspection tools and discussion with inline inspection (ILI) vendors is required in advance. Thus, any anticipated diameter changes require specific in-depth review to ensure that appropriate tool modifications have been made. This term is often used interchangeably with the term dual diameter in the context of inspection tools that are capable of handling a diameter change within a single inspection.

multilayer coating: Refers to the range of external factory applied pipeline coating formulations that uses the deposition of several different films. See *coating* (specifically, Table 16).

multiphase flow: Occurs when two or more distinct phases of matter (gas, liquid, or solid) flow simultaneously through a pipe without physical separation. Immiscible liquids, such as oil and water, would be counted as separate liquid phases. Virtually any combination of gas (or, synonymously, vapor), liquids, and/or solids would constitute a multiphase flow; however, some combinations are more commonly found in pipelines than others. For example, in petroleum pipelines, the term is most often used to refer to the simultaneous flow of oil, natural gas, and water.

The most common reason for transporting fluids via multiphase flow is to move a desirable raw material from its source to a convenient location for processing. In some cases, the desirable material (pulverized mineral ore, for example) is not initially available in a form that can be transported by pipeline, and a multiphase mixture containing the material must be intentionally prepared (for example, by adding water to pulverized coal to form a slurry).

In other cases, the desirable material is initially available as part of a multiphase mixture, often containing several undesirable components as well, but separation into single-phase streams and removal of undesirable components would be too difficult or expensive to carry out at the mixture's source. This would be the case, for example, for a mixture of natural gas and its associated condensate liquids (the desirable components), which might also contain water and sand (the undesirable components).

Table 55 provides examples of gases, liquids, and solids that are commonly transported through pipelines by way of multiphase flows. Transporting multiphase flows in pipelines has important pipeline integrity-related consequences. In general, there are three main integrity-related problems associated with multiphase flow: internal corrosion, erosion, and solids deposition on the pipe wall (which can eventually block the pipeline). Predicting the likelihood and extent of integrity problems in multiphase pipelines can be more difficult than in single-phase systems, because they depend on the physical configuration of the phases in the pipe—whether the individual phases flow as separate stratified layers, as a dispersed mixture, or as some intermediate configuration. This flow configuration, referred to as a flow regime, depends in turn on parameters such as the volume fractions and velocities of the individual phases. Unlike in single-phase systems, flow regimes must often be considered when solving integrity problems in multiphase pipelines. See Table 56.

Table 55. Common examples of multiphase pipeline products (180)

Phase	Examples
Gas/vapor	Natural gas, steam
Liquid	Crude oil, natural gas condensate, water (produced oilfield water, steam condensate), oil/water emulsions
Solid	Produced sand, oil sand tailings, pulverized mineral ores, pulverized coal, petroleum waxes, asphaltenes (from heavy oils), gas hydrates

Table 56. Pipeline integrity problems, and potential solutions, associated with common multiphase systems (180)

System	Examples	Typical Phases	Main Integrity Issues	Possible Solutions
Petroleum production	• Natural gas gathering systems	Natural gas/oil/ water / sand/ petroleum solids	Internal corrosion	• Control velocities to avoid flow regimes where wall is water-wet • Inject corrosion inhibitors
		Oil /water/ emulsion/sand / petroleum solids	Pipe wall erosion	• Remove sand at source • Reduce velocities
	• Crude oil gathering systems		Petroleum solids deposition	• Prevent solids precipitation by controlling temperature or adding precipitation inhibitors; • Maintain sufficiently high velocities to prevent wall deposition; remove solids deposits by frequent maintenance pigging
Wet steam	• Steam distribution systems for heavy oil production	Steam/water	Internal corrosion	• Add oxygen scavengers • Prevent oxygen ingress into system
			Pipe wall erosion due to liquid droplet impingement	• Increase steam quality or superheat steam to minimize liquid dropout • Reduce velocities
Slurries	• Mineral ore slurry pipelines, coal slurry pipelines • Oil sands hydro-transport pipelines, oil sands tailings pipelines	Particulate solids/water Particulate solids/water/ oil	Pipe wall erosion	• Build pipelines above ground, joined with removable couplings • Rotate pipe sections periodically by 90° • Replace pipe sections when wall thickness falls below threshold value
			Solids bed formation	• Maintain flowing velocities above value required to entrain solids • Periodically flush lines to remove solids
			Internal corrosion	• Remove oxygen from water • Provide corrosion allowance in pipe wall thickness

muskeg: Land in a relative cool climate that has poor drainage resulting in the creation of a bog, where the soil tends to be acidic and rich in organic material. The muskeg can be the result of several factors including the presence of bedrock or permafrost at relatively shallow depths. The presence of muskeg is significant from a pipeline integrity point of view because the soil can be quite spongy and susceptible to significant deformation in the presence of relative small loads (i.e., even animals such as moose can become trapped and drown). Thus, pipelines in these areas require special considerations in terms of buoyancy control and potential ground movement associated with freeze/thaw cycles.

N

NACE: See *National Association of Corrosion Engineers.*

narrow axial external corrosion (NAEC): Pitting that occurs on either side of a seam weld or spiral weld often due to *tenting*. This pattern of corrosion may also occur at other pipe orientations (i.e., typically at 3 o'clock and 9 o'clock) beneath wrinkles in the coating caused by soil loading. Because the primary cause of this type of corrosion is from the bridging or tenting of coating over the seam or spiral weld, it is most commonly associated with coatings susceptible to this phenomenon (for example, field applied tape). See *tenting.*

National Association of Corrosion Engineers (NACE): Originally known as "The National Association of Corrosion Engineers" when it was established in 1943 by eleven corrosion engineers in the pipeline industry. These founding members were involved in a regional cathodic protection group formed in the 1930s, when the study of cathodic protection was introduced. Over the course of 60 years, NACE International has become the largest organization in the world focused on the study of corrosion. The association has more than 20,000 members in more than 100 countries across a range of industries including chemical processing, water systems, transportation, and infrastructure.

National Energy Board (NEB) of Canada: An independent federal agency established in 1959 by the Parliament of Canada to regulate international and interprovincial aspects of the oil, gas, and electric utility industries. Mandated by parliament through the Minister of Natural Resources Canada, the NEB exists to promote safety and security, environmental protection, and efficient energy infrastructure and markets in the Canadian public interest.

The NEB has jurisdiction over all international and interprovincial oil and gas pipelines (exceeding 40 km in length). Pipelines that lie entirely within provincial boundaries are regulated by separate provincial regulatory bodies. Prior to building, or adding to, pipelines under NEB jurisdiction, approval from the Board must be obtained. Any application made to the Board may result in public oral or written hearings to review, among other things, its economic, technical, and financial feasibility, and the environmental and socio-economic impact of the project. Further, to ensure that engineering, safety, and environmental requirements are met, the Board audits and inspects the construction and operation of pipelines.

Responsibility for investigating pipeline-related accidents and incidents is jointly shared between the NEB and the Transportation Safety Board (TSB). The roles resulting from joint responsibility are such that the Board is responsible for determining whether its regulations have been followed and the TSB is responsible for investigating the cause and contributing factors. Further, if any key findings indicate that modification of the regulation is in the public's best interest, it is the responsibility of the Board to implement that modification.

natural gas liquid (NGL): Liquid or liquefied hydrocarbons produced in the manufacture, purification, and stabilization of natural gas. The characteristics of various NGLs vary because they comprise hydrocarbons ranging from lighter ends such as ethane, butane, and propane to heavier compounds. NGLs are either blended with refined petroleum products or used directly depending on their characteristics and the specific usage.

NDE: See *non-destructive testing.*

NDT: See *non-destructive testing.*

near-neutral SCC: See *transgranular SCC.*

near-white blast cleaning: Refers to NACE and SSPC standards specifying the nature and degree of surface finish of a metallic component. Specifically, the finish is defined as one from which all oil, grease, dirt, mill scale, rust, corrosion products, oxides, paint, or other foreign matter have been completely removed from the surface except for very light shadows, very slight streaks, or slight discolorations caused by rust stain, mill scale oxides, or light, tight residues of paint or coating that may remain. At least 95% of each square inch of surface area shall be free of all visible residues, and the remainder shall be limited to light discoloration.

NEB: See *National Energy Board.*

NGL: See *natural gas liquids.*

(signal) noise: The portion of the signal in a measurement system that is unwanted and not meaningful. That is, when taking readings, the system used to take the readings will be sensitive to signals that are generated from either external sources or the measurement system itself. Further, the noise can be either random (inconsistent) or systematic (consistent). This concept is relevant in a number of pipeline integrity related situations – particularly in the context of non-destructive testing (e.g., MFL technology, ultrasonic testing) where the noise must be filtered out in order to identify the relevant portion of the signal that can be used to detect and size defects.

nominal pipe size (NPS): One of the two characteristics used to specify pipe in the North American standardized system (the other parameter is either schedule or pipe grade). It essentially refers to the outer diameter of the pipe within manufacturing tolerances (for diameters greater than 12 in.). For pipelines less than 12 in., the actual pipeline diameter is shown in Table 57.

Table 57. Outer pipeline diameter versus NPS (9)

Nominal Pipe Size (NPS)	Outer Pipe Diameter (in.)
3	3.5
4	4.5
6	6.625
8	8.625
10	10.75
12	12.75
14	14

nominal wall thickness (NWT): Refers to the target wall thickness, including corrosion allowance, of a pipeline, recognizing that actual wall thickness will vary within manufacturing tolerances.

non-classical SCC: See *transgranular SCC.*

non-destructive examination (NDE): See *non-destructive testing.*

non-destructive testing (NDT): The group of methods and techniques used to inspect a piece of equipment or materials without impairing its future usefulness. This class of testing encompasses a broad range of technologies including visual, *radiography*, *ultrasonic*, electromagnetic, and dye penetrant testing that vary from industry to industry. This term is used interchangeably with *non-destructive examination*. Table 58 summarizes the most commonly applicable techniques for pipeline integrity. Table 59 shows the relative merits of the various techniques.

Table 58. Summary of pipeline NDT methods [adapted from (1)]

Method	Characteristics Detected	Advantages	Limitations	Examples of Use
Ultrasonics	• Changes in material thickness • Cracks, nonbonds, inclusions, interfaces	• Can penetrate thick materials • Suited to crack detection • Can be automated	• Liquid or gel couplant required • Smooth surface finish needed	• Smart pigging for corrosion and cracking (liquid lines) • In ditch examination of pipe
Radiography	• Changes in density • Voids, inclusions, material variations, placement of internal parts	• Suitable for range of materials and thicknesses • Film remains as permanent record	• Extra precautions needed for handling radioactive material • Limited crack detection abilities	• Weld inspections • Non-surface breaking anomalies
Visual	• Surface characteristics • Finishing, scratches, surfaces breaking cracks, color, corrosion	• No equipment required	• Only applicable to surface or surface opening flaws • Some surface cleaning may be required	• Surface finishing and uniformity • Preliminary assessments of corrosion / cracking mechanisms
Eddy current	• Changes in electrical conductivity • Material variations, cracks, voids, or inclusions	• Readily automated • Moderate cost	• Limited to electrically conductive material • Limited penetration depth	• Crack detection and sizing
Liquid penetrant	• Surface openings • Cracks, surface breaking laminations, porosity, seams or folds	• Inexpensive • Ease of use • Sensitive to small flaws	• Suitable for surface opening flaws only • Not suited to porous or rough surfaces	• In-ditch SCC detection
Magnetic particles	• Leakage in magnetic flux • Surface or near-surface cracks, voids, inclusions, material or geometry changes	• Inexpensive • Sensitive to surface and near-surface flaws	• Limited to ferromagnetic materials • Surface preparation and post survey demagnetization may be required	• In-ditch SCC detection (ferrous material)

Table 59. Relative uses and merits of various NDT methods [adapted from (1)]

Parameter	Ultrasonics	Radiography	Eddy Current	Magnetic Particle	Liquid Penetrant
Capital cost	Medium/high	High	Low/medium	Medium	Low
Consumable cost	Very low	High	Low	Medium	Medium
Time to results	Immediate	Delayed	Immediate	Short delay	Short delay
Geometry effects	Important	Important	Important	Not too important	Not too important
Defect types	Internal	Internal/external	External	External	Surface breaking
Relative sensitivity	High	Medium	High	Low	Low
Formal record	Expensive	Standard	Expensive	Unusual	Unusual
Operator skill	High	High	Medium	Low	Low
Training need	High	High	Medium	Low	Low
Equipment portability	High	Low	Medium/high	Medium/high	High
Dependence on material composition	High	Medium	High	Magnetic only	Low
Ease of automation	Good	Fair	Good	Fair	Fair

normal distribution: The most widely used distribution. It is applied in almost every area where statistics are used. The PDF, $n(x)$, of the normal distribution is given by

$$n(x) = \frac{1}{\sqrt{2\pi}\sigma} e^{-\left(\frac{x-\mu}{\sqrt{2}\sigma}\right)^2}$$

and the CDF, $N(x)$, is given by

$$N(x) = \frac{1}{\sqrt{2\pi}\sigma}\left(1 - erf\left(\frac{x-\mu}{\sqrt{2\pi}\sigma}\right)\right)$$

where μ = location parameter of the distribution
        σ = shape parameter

Table 60 shows selected properties of the normal distribution.

Table 60. Normal distribution properties

Parameters	μ and σ
Domain	$-\infty \le x \le \infty$
Mean	μ
Variance	σ
Mode	μ

Figure 64 shows the shape of the normal distribution for various values of μ and σ.

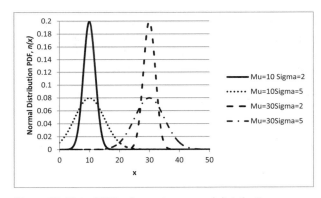

Figure 64. Plot of PDF when using normal distribution.

normal operating pressure: The predicted pressure (i.e., the sum of static head pressure, the pressure required to overcome friction losses, and any backpressure) at any point in a pipe system when the system is operating under a set of predicted steady state conditions. This term is often used interchangeably with *design pressure*.

NPS: See *nominal pipe size*.

NWT: See *nominal wall thickness*.

O

obligate anaerobe: Any organism that does not require oxygen to survive. The most commonly encountered types of obligate anaerobes on pipelines are sulfate-reducing bacteria (SRBs) that can typically tolerate the conditions shown in Table 61.

Table 61. Typical conditions favorable to obligate anaerobes. (18)

Environmental Condition	Range
Temperature	0 to 80 °C
Pressure	Up to 600 bar
Salt	≤ 23%

obstruction: A blockage (either full or partial) of a pipeline due to some physical impediment that restricts the opening. This becomes particularly relevant in the context of in-line inspection where even partial blockage of the pipe can result in tool damage and/or the tool becoming lodged. Examples include *ovality*, collapse, *dents*, undersized valves, *wrinkles*, bends, and foreign objects in the line.

o'clock position: A method used to indicate the circumferential location of a feature on a pipeline. Specifically, superimposing the 12 positions of a clock are on the cross section of the pipeline, facing downstream, to provide a framework for circumferential positioning. See Figure 65. The term is often used interchangeably with *clock position.*

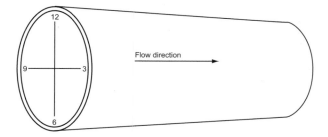

Figure 65. Use of o'clock position to indicate the orientation of an object on the pipeline

OD: See *outer diameter.*

odorant: A substance giving a readily perceptible odor at low concentrations when mixed with a fluid (usually a gas). That is, the odorant is used as a warning sign indicating the presence of the gas before concentrations reach the level of the lower explosive limit. Odorants are most often used in low-pressure distribution gas systems to assist with the identification of potential leaks. In 1937 300 people, most of them children, died in an explosion in a New London, Texas public school building. The natural gas that was being delivered to the building was not odorized. At that time natural gas was odorless, and there wasn't a law on record to mandate odorization. As a direct result of this incident the US Government passed a law that the chemical "mercaptan" be put into natural gas to give it an identifying smell. Mercaptans, also known as thiols, are the most common class of compounds used as odorants.

odorization: See *odorant*.

odometer: An instrument used for measuring the distance traveled by a vehicle. Typically, an odometer is used on in-line inspection tools where the distance traveled is measured by the number of rotations of a wheel. The distance measured this way can be subject to error, especially if there is wheel "slippage" (i.e., lack of rotation) due to excessive debris in the pipeline, which can interfere with the mechanical operation of the wheel.

off potential: Refers to the pipe-to-soil potential reading taken at a given location within the first few seconds after the cathodic protection system is turned off. The reading is taken in millivolts and, where it is not otherwise indicated, refers to the "instant" off reading relative to a standard copper-copper sulfate reference electrode (this is in contrast to the long-term off potential reading where the pipeline is allowed to reach equilibrium with its surroundings). The term is used interchangeably with *polarization potential* and is indicative of the level of protection offered by an impressed current cathodic protection system at a specific location.

off survey: A survey of the pipeline conducted once the cathodic protection system has been turned off for four hours (or more) to ensure that potential gradients have stabilized.

Office of Pipeline Safety (OPS): In the United States, the Department of Transportation's (DOT) Pipeline and Hazardous Material Safety Administration (PHMSA), acting through the Office of Pipeline Safety (OPS), administers the Department's national regulatory program to assure the safe transportation of natural gas, petroleum, and other hazardous materials by pipeline. OPS develops regulations and other approaches to risk management to assure safety in design, construction, testing, operation, maintenance, and emergency response of pipeline facilities.

off-line inspection: Refers to inspection of a pipeline with a smart pig where the pipeline must be taken out of service. Several factors could result in the need for taking the line out of service; these include, but are not limited to:

- a lack of launching and receiving facilities
- product flow speeds that are too high/low during normal operations
- the need for a specific type of fluid to accommodate the inspection technology (e.g., a liquid couplant (vs., gas) is often required for ultrasonic technology)
- mechanical constraints in the pipelines (such as unpiggable valves or unbarred tee junctions)

ohmic drop: Refers to the potential drop that can be measured between the reference and working electrodes due to the resistance of the electrical circuit itself. In the context of cathodic protection systems on the pipeline, there are three major components: IR drop between the reference cell and the soil, the soil itself, and the polarization at the pipe coating and soil interface.

oil line: A pipeline where the primary product is a hydrocarbon that, under atmospheric conditions, exists as a liquid and has low vapour pressure.

on-line inspection: Refers to the inspection of a pipeline with a smart pig where the pipeline remains in service with minimal (or insignificant) disruption to normal operations. *See in-line inspection.*

on potential: Refers to the potential reading at a given location on the pipeline with the cathodic protection system operating. The readings are always taken relative to a reference electrode – typically a copper-copper sulfate electrode placed directly above the pipeline. The typical criterion for ensuring that the pipeline is sufficiently protected is to achieve a -0.85 mV value. See also *cathodic protection*.

on survey: The group of potential readings taken along a pipeline with the cathodic protection system turned on. The readings are always taken relative to a reference electrode – typically a copper sulfate electrode. The purpose is typically to determine the effectiveness of the cathodic protection system and to determine if any maintenance or repair work is required.

one-call system: A system set up to establish a common link between all parties that may be affected by any type of excavation (i.e., either individuals or companies) and those who operate underground facilities. The goal of these systems is to reduce injury and damage as a result of unauthorized excavations by providing a single point of contact for all inquiries. Systems are typically funded either by the government or organizations responsible for operating underground facilities associated with communications, gas distribution, gas transmission/gathering, electric power, product pipelines, water, and wastewater lines.

on-stream: A term used interchangeably with *in-service* and *on-line,* referring to the state of a pipeline system as being fully operational. See *in-service.*

open-circuit potential: A potential that is measured on a cathodic protection circuit when there is no current flowing. Specifically, the open-circuit potential of various anodes (or corroding surfaces) will differ based on their properties (relative to a standard reference cell) and must be accounted for in the basic design of the cathodic protection system. The term is often used interchangeably with *corrosion potential.*

operating company: The organization responsible for the daily operations of moving product from one location to another through a pipeline system. Typically, this would also involve long-term maintenance planning and capital expansion work. This is not always the same organization that owns the asset.

operating conditions: Refers to the typical range of circumstances a pipeline is subjected too during routine service. For example, operating conditions would include the temperature (internal fluid temperature as well as environmental temperature), pressures, and soil loadings that a pipeline is subjected to during the course of routine operations.

operating pressure: Refers to the typical maximum pressure that a pipeline experiences during the course of routine operation.

operator: 1. An individual who engages in the operation of facilities used in the transportation of gas, liquids, or other products through a pipeline. **2.** See *operating company*.

Operator Qualification Rule (OQ Rule): A set of regulatory requirements instituted by the US Department of Transportation's (DOT) Office of Pipeline Safety in 1999. The regulations, a part of 49 CFR 192 and CFR 195, require operators to develop a written qualification program to evaluate their employees and contractors in performing covered tasks. Specifically, individuals must be able to recognize and respond to abnormal operating conditions that may be encountered while undertaking covered tasks. In keeping with a performance-based approach, the regulation provides guidance regarding what tasks are "covered"; however, final determination of which activities are "covered" remains the responsibility of the operator. The guidance for a "covered task" consists of four "tests" that the task:

1. Is performed on a pipeline facility
2. Is an operations or maintenance task
3. Is performed as a requirement of the regulation
4. Affects the operation or integrity of the pipeline

Further guidance on the regulation can be found in a consensus document entitled *API 1161 – Guidance Document for the Qualification of Liquid Pipeline Personnel* published by the American Petroleum Institute, and in ASME B31Q.

OPS: See Office of Pipeline Safety.

optical imagery: A satellite-based technique used to identify geotechnical hazards. Specifically, repeated views of the same location can be used to identify ground, slope, and water course movement over time from a remote location; however, movements must be reasonably large to be identified through this method. See also: encroachment.

OQ Rule: See Operator Qualification Rule.

organic anionic inhibitor: A class of anionic inhibitors that are composed mostly of carbon and hydrogen. See the entry for inhibitor for a more detailed discussion regarding the factors affecting the effectiveness of inhibitors as well as mechanisms associated with them.

organic cationic inhibitor: A class of cationic inhibitors composed mostly of carbon and hydrogen. See the entry for inhibitor for a more detailed discussion regarding the factors affecting the effectiveness of inhibitors as well as mechanisms associated with them.

organic inhibitor: A class of compounds, composed mainly of carbon and hydrogen, used in pipeline applications to reduce internal corrosion. See the entry for inhibitor for a more detailed discussion regarding the factors affecting the effectiveness of inhibitors as well as mechanisms associated with them.

orientation: The position of an object (e.g., fitting) or flaw (e.g., in-line inspection anomaly) on or in the pipeline. Circumferential orientation typically is expressed as an o'clock position or in degrees. See o'clock position.

outage: A situation on a pipeline, or a section of a pipeline, that has been taken out of service. The term is most often used in relation to modifying existing facilities (e.g., repair and/or upgrade) or tying-in new facilities where the nature of the work requires a disruption in service either for safety reasons or because the pipeline product must be evacuated for work to proceed.

outer diameter (OD): The actual diameter of the outside of the pipeline wall. See Figure 66. It should be recognized that the measurement will vary slightly over the length of the pipeline depending upon the manufacturing tolerances of the pipe.

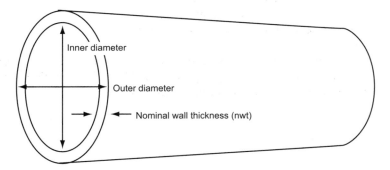

Figure 66. Outer diameter on a pipeline.

ovality: An indicator of the degree of "out-of-roundness" of a pipe section. Specifically, the term refers to a situation where the circular pipe deforms into an ellipse, usually as the result of external forces.

over-pressure: A situation where the (non-routine) operation of a pipeline results in the internal pressure exceeding the *maximum allowable operating pressure*. This is generally tolerated for short-duration upsets.

overprotection: The amount of current provided through a cathodic protection system that is above and beyond what is necessary to meet cathodic protection criteria. Because current distribution is not perfectly uniform across a pipeline system, overprotection will occur in some areas along the pipe to achieve minimum levels of protection for the full length of the pipeline. Significant and sustained levels of overprotection can lead to coating damage and must be considered as part of the cathodic protection system design and implementation.

owner-inspector: An individual who monitors the work of a contractor with a view to representing the owner's interests. The role is essentially one of quality control, and individuals are often qualified based on the results of some form of examination that will depend on the specifics of the work to be monitored. Most companies have well-defined policies and procedures indicating when inspection (either internal or third party) is required. The term is often shortened to the term *inspector*.

oxidation: A generic term for a chemical reaction where the atoms of an element lose electrons. More specifically, oxidation is the first part of a corrosion reaction where the iron atoms in the pipe material are compromised by losing electrons. See *corrosion*.

oxygen (O_2): A gaseous compound, which is present in the atmosphere, soil, and in many industrial processes. It is present in our atmosphere forming ~21% (by volume) of our environment and is essential for most biological processes. Oxygen is often introduced into gas and liquid pipelines – most commonly through the presence of water (H_2O). Oxygen can often participate in and accelerate corrosion reactions – especially microbiological activity. See *microbiological corrosion*.

P

P & ID: See *piping and instrumentation drawing.*

paint: A liquid to paste-like coating material that is applied predominantly by spraying, brushing or rolling – it is often one of the first layers to be applied as part of a coating system. The term is often also used interchangeably with the more specific term primer.

paint system: A combination of coating material, usually in a liquid or paste form, applied in layers to pipe to provide surface protection. Each layer is formulated to have a specific function within the system. For example, the first layer is likely to be specially formulated for dealing with surface preparation and ensuring good adhesion for subsequent layers for color, environmental protection, etc.

parent material: The original (i.e., unaltered) pipe metal that may have different properties from the metal deposited through the welding process.

partially turbulent flow: Typically refers to single-phase flow where the flow is neither fully turbulent or fully laminar. Specifically, the flow remains laminar along the pipe surface and transitions to fully turbulent flow in the center of the pipe cross section. See Figure 67.

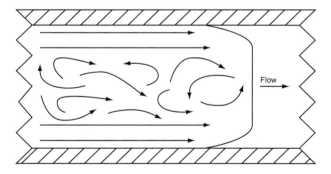

Figure 67. Partially turbulent flow. (180)

particulate erosion: The damage resulting from very small fragments of material (e.g., grit, soil, metal filings) colliding with the internal pipe wall at relatively high velocities. The locations most susceptible to this type of damage are points where there is a change in the direction of flow (i.e., bends) or where equipment impinges into the flow (e.g., orifice meters).

parts per million (ppm): A dimensionless (i.e., no units) quantity used to describe the presence of a substance existing in (relatively low) concentrations. Similar to the concept of a percentage (where a value is expressed as a proportion to a denominator of 100), ppm also represents a proportion; however in this case, the denominator is one million. The concept is used to describe the concentration of substances in all three phases (i.e., liquids, gases, and solids).

When describing a solution, parts per million can sometimes refer to concentrations in grams per gram, volume per volume, or moles per mole.

passivating inhibitor: A class of inhibitors that encourages the formation of an oxide layer (*passive film*) on the pipe surface. This oxide layer is a physical barrier that renders the pipe chemically inert and limits further corrosion. See *inhibitor* for a more detailed discussion regarding the factors affecting the effectiveness of inhibitors and mechanisms associated with them.

passivation: A process where a metallic surface is rendered inert through the creation of a physical barrier (i.e., usually an oxide layer) that is created through a corrosion reaction.

passive film: A physical barrier (i.e., usually an oxide layer that is a natural by-product of the corrosion process) that renders the pipe surface chemically inert. The barrier then prevents any further corrosion in the area unless the integrity of the film is compromised (through impact of a small particle for example) – at which point, the process repeats itself.

patrolling: A method for monitoring the conditions of the pipeline's *ROW*. Specifically, the objective is to observe the *ROW* surface conditions to identify leaks, unauthorized construction activity and other factors affecting safety and operation. The frequency of patrols is determined by the size of the pipeline, operating pressures, class location, terrain, weather and other factors that could impact the integrity of the pipeline. Methods for patrolling include walking, driving, flying, or other means of traversing the right of way.

patina: The color change that results from corrosion product forming a fine coating on the surface of metal alloys. The mostly commonly observed example is a greenish oxide on copper and copper alloys that have been exposed to the atmosphere.

PDF: See *probability density function.*

Pearson survey: An electromagnetic technique used to detect below-ground coating defects on a pipeline system. Specifically, an AC signal is injected onto the pipeline, and the potential gradient is measured between two portable ground contacts. The gradient is measured and recorded along the length of the pipeline in this manner, and coating defects are identified through localized increases in potential gradient. Figure 68 shows the principle of a Pearson survey. It should be noted that DCVG techniques are more commonly used due to their significant technical and cost benefits.

pencil probe: A long, thin transducer and its housing used for taking measurements in difficult-to access areas of equipment. Specifically in the context of pipeline integrity, this term is most commonly used to describe an ultrasonic probe for taking wall thickness measurements of corroded pipe, especially in areas where pitting corrosion is present.

penetrant: Refers to a class of liquids used in visual inspections of the pipe surface. Penetrants are formulated to easily flow into surface breaking defects to highlight them through contrast relative to uncompromised material. Table 62 shows five main types. See *liquid penetrant inspection* for additional detail on the non-destructive testing technique.

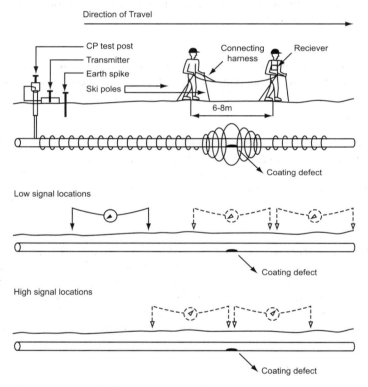

Figure 68. Principle of Pearson survey [adapted from (195)]

Table 62. Key characteristics of various types of liquid penetrant (1)

Penetrant Type	Key Characteristics
Dual Purpose	Produces both fluorescent and color contrast visible indications
Fluorescent	Emits visible radiation when excited by black light
Solvent-removable	Traces removable wiping material lightly moistened with solvent remover
Visible	Characterized by an intense color, usually red
Water-washable	With a built-in emulsifier

Performance-based regulation: A philosophical approach to rulemaking that focuses on outcomes and targets for operating and maintenance programs – with significantly less emphasis on the specific approach an operator may use to achieve the desired outcomes. This is in contrast to *prescriptive regulation*.

performance specification: A set of criteria that one must meet at the point of completing a pre-defined scope of work. This is most often used in the context of a regulator setting performance criteria for a pipeline operator (for example, in the area of safety) or an owner prescribing the requirements for contractors undertaking work (for example, in installing a piece of equipment). The key implication of this approach is that the party setting the performance specification allows the organization undertaking the work to exercise their best judgment in determining how best to meet the performance specification – as such, one could expect a wide range of methods to be used in meeting any given performance specification.

permafrost: Any rock or soil material that remains frozen and does not thaw seasonally (i.e., has remained below 0°C continuously for two or more years). A thin top layer, defined as the active layer, may seasonally thaw and freeze and is often sufficient to support basic plant life. Permafrost is encountered at higher latitudes – or at high altitude in lower latitudes. Depending on the type of soil and bedrock, ice is not always present. Permafrost is of particular concern to pipeline design and integrity due to the potential for differential thaw settlement, frost heave at transitions from frozen to unfrozen terrain, and the stability of thawing permafrost slopes. Pipelines that operate at temperatures that result in thawing, particularly in ice rich areas, may face issues associated with buoyancy and may be subjected to significant stress/strain due to ground movement. Table 60 shows three main types of permafrost.

permitting: The process of applying for and receiving approvals to commence work from all relevant regulatory authorities. The specific requirements obviously vary based on the nature of the activity as well as the jurisdiction in which the work is being done. The term is most often used in the context of new construction or maintenance work, especially if there are potential environmental implications.

phased array: A specific configuration of automated ultrasonic testing where multiple pulse generators (typically between 4 and 32) are individually wired and pulsed at the same frequency, focal length, and incidence angle; the defining factor is that they are time shifted. This is a relatively new technology and therefore still less cost effective when compared to multiple conventional transducers. Benefits, relative to the more traditional transducers, include scanning speed, flexibility in setups and incidence angles, and a smaller footprint.

Table 63. Key characteristics of various types of permafrost (26)

Type of Permafrost	Key Characteristics
Cold	Typically defined as an annual average ground temperature < -2°C Tolerates the introduction of considerable heat without thawing
Warm	Typically defined as an annual average ground temperature $\geq -2°C \leq 2°C$ Addition of small amounts of heat may induce thawing
Discontinuous	An area in which there co-exist pockets of permafrost and unfrozen soil; the percentage of each varies depending upon the location.

PHMSA (Pipeline and Hazardous Materials Safety Administration): One of ten agencies within the US Department of Transportation (DOT), PHMSA's role is to protect the American public and the environment by ensuring the safe and secure movement of hazardous materials to industry and consumers by all transportation modes, including pipelines. PHMSA was created under the Norman Y. Mineta Research and Special Programs Improvement Act (P.L. 108-426) of 2004, which was signed into law by President Bush on November 20, 2004. The creation of PHMSA provides the Department a modal administration focused solely on its pipeline and hazardous materials transportation programs. Through PHMSA, the Department develops and enforces regulations for about 2.3 million-mile pipeline transportation system.

pickling: A range of electrolysis and chemical treatments used for surface cleaning of metal components. The most common use in pipeline applications is the removal of mill scale, rust, and rust scale with the goal of ensuring that the resulting surface is free of all scale, rust, unreacted chemicals (from the treatment), or any other foreign matter.

pig: A class of devices that are inserted into and travel throughout a length of pipeline, driven by the product flow. As such, they require temporary *pig traps* (i.e., *launcher/receiver* facilities) to be used. They fall into two categories: "utility" pigs - which perform a function such as cleaning, separating, or dewatering the pipeline; and *in-line inspection tools* (sometimes referred to as "intelligent pigs" or "smart pigs"), which provide information on the condition of the line as well as the extent and location of any problems. See Table 64.

pig signal: A device used to indicate when the pig has reached a certain point in the pipeline. This is usually achieved by attaching a triggering device or "signaller." This may be activated by the pig physically moving a lever or plunger that protrudes into the line (referred to as intrusive) or by remotely sensing the pig's presence from outside the pipe wall by, for example, a change in the magnetic field. This is usually referred to as a non-intrusive signaller.

pig trap(s): The set of facilities used launch and receive pigs at either end of the pipeline segment to be pigged. These facilities can be either permanent or temporary depending on the specifics of the situation. See Figure 69. While there can be specific requirements resulting from the tool/technology used, Table 65 shows key considerations.

Table 64. Major pig types and typical uses [adapted from (10)]

Major Pig Type	Subtype	Description	Primary Usage
Utility tools	Foam	• Most basic and inexpensive of utility pigs • Bullet-shaped, lightweight and easy to work with • Flexibility allows them to negotiate a variety of pipe diameters and geometries (e.g., tight radius, mitre bends) • Range of foam densities (can therefore be very soft to very hard)	• Drying (bare configuration) • Cleaning (with metal or plastic bristles) • Batching (separation of fluids) • Product removal
	Spherical	• Oldest and most widely used utility pig • Usually polyurethane globes, filled or inflated with liquid (usually water, glycol, or lightweight oil) • Can be solid spheres • Requires all tees to be barred so pigs do not become lodged in pipeline	• Heavy-duty service in the removal of liquid hydrocarbons and water from pipelines • Batching (separation of fluids) • Hydrostatic testing
	Mandrel	• Most versatile of utility pig • Usually fabricated from a metallic central body tube ("mandrel") with a combination of various attachments (e.g., cups, brushes, discs etc.,) suited to the application • Sealing elements can be cups or discs (bi-directional pig) • Ability to negotiate range of pipe geometries is a function of the specific pig configuration (i.e., attachments used etc.,)	• Batching (separation of fluids) • Gauging • Sealing • Cleaning / brushing • Confirmation of minimum internal pipeline diameter (with use of gauge plates)
	Solid cast	• General purpose utility pig molded in one piece, usually from polyurethane • Ability to attach various components (e.g., brushes) for specific services • Range of configurations (i.e., combination of discs, cups etc.,) • Easy to handle and with good sealing capabilities • Tends to experience high wear and as a result are more common in small diameters (small diameter lines are usually short) but can be obtained up to 36-in. diameter	• Batching (separation of fluids) • Displacement • Cleaning
	Gel (jelly)	• A "gel" pig is a semi-solid chemical substance (i.e., highly viscous) injected or placed inside a pipeline – usually in a pig train • Gels can be water-based, or use a range of chemicals, solvents, and even acids	• Debris pickup/cleaning (primarily for powder like debris such as iron sulfides and rust) • Water and condensate removal • Batching (separation of fluids)

Table 64 *continued*

Major Pig Type	Subtype	Description	Primary Usage
In-line inspection tools	Metal Loss	• Range of technologies used including: magnetic flux leakage, ultrasonics (primarily liquid lines) • Range of pipe diameters available • Requires a pre-run review to ensure safe tool passage through specific pipeline geometries • Typically requires cleaning runs and confirmation of pipeline internal diameter in advance of metal loss inspection • Can be tethered for small diameters/lengths	• Detection and sizing of corrosion, mechanical damage and other defects such as mill anomalies • MFL technology can be adapted to identify changes in pipeline properties (i.e., hard spot detection)
	Geometry	• Range of sophistication technologies used including basic caliper tool to inertial mapping systems providing detailed 3D dimensional pipeline profile information • Range of pipe diameters available • Requires a pre-run review to ensure safe tool passage through specific pipeline geometries • Typically requires cleaning runs and some indication of pipeline internal diameter in advance of metal loss inspection	• Detection of diameter restrictions • Bend measurement • Pipeline mapping • Mechanical damage/strain measurement
	Crack detection	• Range of technologies used including transverse MFL, ultrasonics, EMAT • Range of pipe diameters available • Requires a pre-run review to ensure safe tool passage through specific pipeline geometries • Typically requires cleaning runs and confirmation of pipeline internal diameter in advance of metal loss inspection	• Detection and sizing of crack and crack like anomalies
	Pipeline sampling	• Typically, a mandrel pig mounted with sensors appropriate to the required application • Requires a pre-run review to ensure safe tool passage through specific pipeline geometries	• Leak detection • Temperature and pressure record mapping • Product sampling • Wax deposit measurement
	Visual inspection	• Usually, a tethered tool suited to inspection of specific pipeline location • Requires a pre-run review to ensure safe tool passage through specific pipeline geometries • Typically requires cleaning runs and confirmation of pipeline internal diameter in advance of metal loss inspection	• Internal visual inspection of defects, pipeline geometry etc.

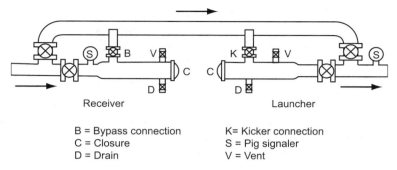

B = Bypass connection   K= Kicker connection
C = Closure   S = Pig signaler
D = Drain   V = Vent

Figure 69. Typical pig launcher/receiver facilities.

Table 65. Key elements of pig trap design (124)

Trap Characteristic	Launcher Considerations	Receiver Considerations
Overbore	• Should be short enough to allow ease of tool loading (specific to vendor equipment) • Usually 2 diameter sizes larger than NPS	• Largely a function of the tool length • Should be the length of the tool at a minimum for gas lines to ensure tool stoppage without hitting door • For liquid lines, should be the tool length or 3 ft (as a minimum) • Usually 2 diameter sizes larger than NPS
Nominal bore	• Combined length of overbore and nominal bore pipe should be sufficient to fully load tool and close trap door	• Largely a function of the tool length • Should be at least the length of the tool trailing modules to ensure that tool does not block isolation valve when it loses drive at the overbore
Kicker line	• Position not critical – ideally located at 3 or 9 o'clock position near trap door	• Bypass line should be located as close to the reducer as possible
Reducer	• Eccentric preferable for ease of tool loading	• Eccentric preferable to reduce sudden velocity changes upon receive
Balance Line	Not essential depending on tool	
Drain-line	Drain-lines can help with waste containment – even for gas lines; best located near trap door to assist with clean-up of barrel	
Access area	Access area should be sufficient to allow safe working space for pig, support equipment (such as pig tray and crane) as well as personnel	
Vent	To the extent possible, best positioned such that vent is not coincident with sensors on smart tools to prevent damage during purging	
S-Bend	Minimum 3-diameter radius of bends is preferable – particularly in thick wall pipe associated with launch/receive facilities	

piggable: A specific pipeline section that is suitable/ready for pigging operations – for either cleaning, gauging, or inspection – is considered piggable. Specifically, the term refers to the physical characteristics of the pipeline section and whether an inspection tool could travel through the section undamaged while successfully collecting data. See *unpiggable*.

pigging procedure: A series of steps, tasks, and safety precautions taken to ensure a successful pig run. While each situation is unique and requires a detailed assessment of the specific conditions involved, Table 66 shows some of the key elements of a typical pigging procedure.

pinhole leak: A pinhole leak is the release of pipeline product from a defect that has a width and length less than the equivalent of one wall thickness of the pipe. See *defect classification*.

pinhole: 1. Refers to a coating defect where broken blisters appear on the surface of the pipe. In extreme cases, this may appear as a general roughness of the coating. The defects are most often caused by inadequate surface preparation whereby water and other contaminants are driven off the pipe during the heating and application process creating bubbles in the coating. (14). **2.** A defect that is not usually detected at the pipe mill and often is discovered after construction and commissioning of the pipeline. This is because the pinhole defect can be plugged by mud while in the ditch. **3.** Also refers to the approximate shape of a highly localized area of metal loss defect as identified through in-line inspection. Specifically, the dimensions of the anomaly are

$$width < A$$

and

$$length < A$$

where A = wall thickness if $t < 10$mm, else $A = 10$mm

The definition is based on a document published by the *Pipeline Operators Forum*; additional detail is provided in Figure 29 under *defect classification*.

pinwheel pig: A pig used for aggressive pipeline cleaning where hardened metallic pins are mounted on a body (typically constructed from a material such as urethane). The pins are primarily to loosen debris from the internal pipe surface. The degree of aggressiveness can be specified for the circumstance by varying the number and length of the pins as well as the hardness of the pig body. Depending on the specific nature of the cleaning operations, the pinwheel pig may need to be coupled with additional tools to ensure sufficient debris removal occurs and/or differential pressure to keep the tool moving.

pipe: A hollow cylinder, typically made of metal, plastic, or composite, used to transport liquids, gases, or slurries from one location to another. See *pipeline*.

pipe grade: See *grade*.

Table 66. Key elements of a procedure used for pipeline pigging (124)

Item	Description
Safety	Addresses key considerations for handling the tool, any hazardous materials as well as specific operational constraints
Launch / Receive	Ensures that line conditions are appropriate for a safe tool launch/ receive and ensures service disruptions are minimal
Tracking	Identifies key tracking locations, coordinates, access, and approximate timing to ensure safety of personnel (especially in remote locations)
Waste Handling at Receiver	Ensures that any pipeline product, debris, and any other hazardous materials are handled in an appropriate manner. (For example, iron sulfide, which can be contained within pipeline debris, is extremely flammable when exposed to atmospheric conditions.
Speed Control	Identifies operational conditions required (i.e., pressure, temperature, and flow rates) required for maintaining tool speeds within the optimal range for pig performance
Situation Specific Considerations	Any run specific situations that require coordination and planning, such as section isolation, customer service interruption, access to very remote locations (e.g., helicopter access) etc.,

pipe integrity: Refers to the ongoing safe operation of a pipeline system without disruption to service, the surrounding environment, or individuals working or residing in its vicinity.

pipe locator: A device used to locate underground pipe without excavation. The technologies used to accomplish this vary, especially depending on the pipe material. However, the most commonly used techniques rely on sensing changes in magnetism or dielectric properties (resulting from the presence of an underground pipe). Techniques such as ground penetrating radar are also used.

pipe mill feature: Any (generic) anomaly that results from the fabrication process. As such, the feature may/may not pose an integrity threat and must be evaluated on a case by case basis. See *mill-related anomalies*.

pipe number: A unique identification code assigned by an operating company to each pipe segment based on an internal classification system. This is often critical to internal data management systems that may be as simple physical file management of paper drawings or as complex as a fully integrated GIS system integrating all key pipeline parameter.

pipe schedule: One of the key parameters for defining the specifications of line pipe – specifically a reference to standard pipe *(nominal) wall thicknesses*.

pipe segment: Usually, a length of pipe bounded by changes in pipeline attributes that result in a material change to risk levels compared to adjacent segments. The change in risk level can be driven by a change in any number of factors including, but not limited to wall thickness, pipe grade, soil type, coating type, population density, defect density/ severity and product flow characteristics. A pipeline segment can vary in length from a few joints to tens of kilometers.

pipe tally: Following an in-line inspection, the vendor will provide a listing of all pipeline component and anomalies in the form of a pipe tally. While individual operators can (and do) modify the reporting requirements associated with this listing. Table 67 shows what is typically contained.

pipe to electrolyte potential: The voltage difference between the pipe surface and any electrolyte that may be present in its vicinity (e.g., under a coating defect). The voltage would be indicative of the potential for galvanic corrosion in that specific location.

pipe to soil interface: Refers to the area where the pipe transitions from being located above ground to being located below ground. This can be a particular area of concern from an integrity point of view because this transition area is subject to a (relatively) large amount of variability in environmental conditions. Specifically, varying moisture levels, soil stresses, and higher temperature differentials could lead to increased susceptibility to time dependent failure mechanisms.

pipe-to-soil potential: The voltage difference between the pipe surface and the surrounding soil. The voltage would be indicative of the potential for galvanic corrosion and/or effectiveness of the cathodic protection system in that specific location. It should be noted that because soil does not provide a "stable" reference point, the term is a slight misnomer as the measurement is actually taken relative to a reference cell (most commonly a copper-copper sulfate reference cell).

Table 67. Key elements of a "pipe tally" typically provided by in-line inspection vendors (124)

Item	Description
Log distance	The distance traveled by the tool from the point of launch – typically measured by an onboard odometer. This measurement is prone to error from slippage and other mechanical sources of error and should be adjusted using known reference points on the pipeline when locating anomalies.
Girth weld number	Refers to the numbering of each girth weld as an aid to locating the feature both in the field as well as between multiple runs.
Upstream weld distance	The distance traveled by the tool from the closest upstream weld as measured by the onboard odometer.
Joint length	Th e length of the section of pipe in which the anomaly or component is located (i.e., the distance between the closest upstream and closest downstream girth welds).
Feature type	While specific terminology/abbreviations vary by vendor, this field indicates whether the item is an anomaly, girth weld, above-ground marker, or other pipeline attribute detected by the tool.
Feature identification	A more specific description of the feature identified in the previous column. For example, a girth weld may be associated with a wall thickness change or an anomaly maybe identifiable as a dent (or a dent with corrosion). Again, specific abbreviations/naming conventions will vary by vendor.
Anomaly dimension classification	A classification system for corrosion based on its dimensions (primarily length and width).
Clock position	The circumferential location of a feature on a pipeline using the twelve positions of a clock, superimposed on the cross section of the pipeline, facing downstream.
Nominal wall thickness	The target wall thickness, including corrosion allowance, of a pipeline recognizing that actual wall thickness will vary within manufacturing tolerances of each joint or pipeline component as measured by the inspection tool.
Reference wall thickness	Only relevant for ultrasonic tools where the actual wall thickness of each pipe joint, as measured by the tool, is reported. Where there is a variation in the reference wall thickness over the length of the joint, the most frequently measured reference wall thickness is typically provided. If agreed by client and contractor the minimum or average wall thickness can also be given as reference wall thickness.
Width of anomaly/feature**	The width of the reported feature
d/t	A calculation of the depth of the indication as a percentage of the wall thickness.
Surface location	A reporting of whether the feature is internal, external, mid wall, or undetermined.
Calculated burst pressure	A reporting of the calculated failure pressure (this may be divided by the design pressure).

pipeline: Comprises all parts of the physical facility through which slurries, liquids (crude oil, petroleum products), or gases (natural gas, carbon dioxide) are transported including, but not limited to, pipe, valves and other equipment attached to the pipe, compressor units, pump stations, metering stations, regulator stations, delivery stations, breakout tanks, holders, appurtenances connected to line pipe, and fabricated assemblies. See Figure 70 for a diagram pipeline systems. Table 68 shows the major types of pipeline, and their general usage.

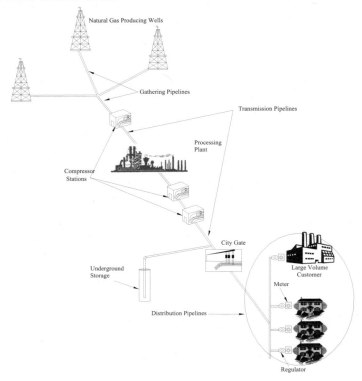

Figure 70. Schematic of various pipeline systems. (199)

Pipeline and Hazardous Materials Safety Administration: See PHMSA.

pipeline construction specification: The set of standards developed by an owner company documenting the practices and requirements that guide all construction practices on a given pipeline system. Specifications can encompass a wide range of topics including, but not limited to, material selection, construction techniques, and minimum design standards.

pipeline facility: See *pipeline*.

pipeline lifecycle: The time period that encapsulates the construction, operation as well as retirement/abandonment of a pipeline system. This term is most often used with reference to a design philosophy that is holistic in the sense that is goes beyond considering just the operational lifecycle of the pipeline; it also addresses retirement/abandonment issues once the system has reached the end of its useful life.

pipeline monitoring methods: The numerous techniques to monitor and assess the state of the pipeline and its surroundings; an operator should assess the greatest damage/defect risk to the pipeline, then select a monitoring/inspection method to reduce that risk. Table 69 shows some of the more common methods.

Pipeline Operator's Forum (POF): An international non-profit, informal group enabling pipeline integrity engineers to share and build best practice, with the goal of raising the standard of pipeline integrity management. It has more than 20 members including several multinational corporations.

Table 68. Three main types of pipelines: distribution, gathering and transmission (8)

Type	Description
Distribution line	A pipeline other than a gathering or transmission line. Within this grouping, there are two main types of lines: A "main," which is a distribution line that serves as a common source of supply for more than one service line. A "service line," which is a distribution line that transports product from a common source of supply to an individual customer or to multiple residential or small commercial customers served through a meter header or manifold. A service line ends at the outlet of the customer meter or at the connection to a customer's piping, whichever is further downstream, or at the connection to customer piping if there is no meter.
Gathering line	A pipeline that transports product from a current production facility to a transmission line or main.
Transmission line	A pipeline, other than a gathering line, that: Transports product from a gathering line or storage facility to a distribution center, storage facility, or large volume customer that is not down-stream from a distribution center; Operates at a hoop stress of 20 %or more of SMYS; or Transports product within a storage field.

Table 69. Summary of common pipeline monitoring methods and their applications [adapted from (8)]

Defect/ damage type	Satellite/ aerial / ground survey	Intelligent pigging	Product quality	Geotechnical survey & strain gauge	CP & coating survey	Hydrostatic testing
Third party damage	Proactive	Reactive				Reactive
External corrosion		Reactive			Proactive	Reactive
Internal corrosion		Reactive	Proactive			Reactive
Cracking		Reactive				Reactive
Coating damage					Proactive	
Material / construction defect		Reactive				Reactive
Ground movement		Reactive		Reactive		

Pipeline Research Council International (PRCI): An international, US-based, industry consortium that was formed in 1952. The association has more than 50 members and is primarily composed of pipeline operators, but also includes several service providers and equipment manufacturers. The organization is set up to ensure that research and development is undertaken as efficiently as possible through information sharing, cooperative project development, and broad dissemination and application of results (to member companies). PRCI undertakes research in a range of areas concerning the pipeline industry including operations, maintenance, and regulatory issues.

Pipeline Safety Act: The US law that authorizes the Secretary of Transportation to prescribe safety standards for the transportation of gas and hazardous liquids. Table 70 shows the relevant acts that cover various aspects of the pipeline industry.

pipeline system: See *pipeline.*

piping & instrumentation drawing (P&ID): A drawing or set of drawings that details the process flow, measurement, and control system layout for a facility. Specifically, it details how equipment in the process interconnects with the instrumentation and controls that regulate the process.

pit gauge: An instrument used to directly (mechanically) measure the depth of pitting corrosion; specifically, the instrument has a long, thin probe than can be inserted into areas of highly localized corrosion such that depth, relative to the surrounding pipe surface, can be measured.

pitting: The approximate shape of localized metal loss defect as identified through in-line inspection. Specifically, the anomalies represent corrosion that is more localized then generalized than general corrosion and less localized than pitting as well as roughly circular in nature. The definition is based on a document published by the *Pipeline Operators Forum*; see additional detail and Figure 29 under *defect classification.*

Table 70. Key elements of US *Pipeline Safety Act* (182)

Part #	Title
Part 190	Enforcement Procedures
Part 191	Reporting Requirements
Part 192	Transportation of Natural and Other Gas by Pipeline: Minimum Federal Safety Standards
Part 193	LNG Facility Safety
Part 194	Response Plans for Onshore Oil Pipelines
Part 195	Liquid Pipeline Minimum Safety Standards
Part 198	Grants to State Pipeline Programs
Part 199	Drug Testing
Part 40	Procedures for Workplace Drug Testing Programs (OST)

Pitting resistance equivalent numbers (PREN): A theoretical method that allows the comparison of the pitting corrosion resistance of steels. While there are a number of equations that can be used, the most common form is based on the chromium, molybdenum, tungsten and nitrogen content of the steel as follows:

$$PREN = Cr + 3.3(Mo + 0.5W) + 16N$$

where Cr = chromium content (%)
       Mo = molybdenum content (%)
       W = tungsten content (%)
       N = nitrogen content (%)

The PREN value is often an additional specification (i.e., above and beyond compositions implied by ASTM standards) for steels used in corrosive environments (such has those containing hydrogen sulfide). See also *duplex steel* and *super duplex steel*. (188)

plastic deformation: The change in a component's shape that occurs once stresses exceed the yield stress of the material, i.e., the deformation is permanent and the material does not return to its original shape once the loads are removed. See *tensile strength* for a detailed discussion of this topic.

platinum anode: One of the most common types of anodes used in impressed current cathodic protection systems.

plug valve: A specific type of valve used primarily for isolation; however, specialized designs can be used for pressure or flow control in pipeline applications. Plug valves can be either full port (i.e., have an opening that is the full diameter of the pipeline – or greater) or reduced port (i.e., have an opening that is the less than the full diameter of the pipeline), depending on the application. Older main-line plug valves can pose a particular challenge for the safe passage of in-line inspection tools. Figure 71 shows the basic form of a plug valve.

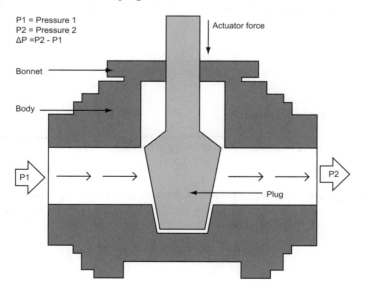

Figure 71. Principle of plug valve operation.

POD: Probability of detection. See *detection.*

PODS™ (Pipeline Open Database Standard): An independent database model for gas and liquid gathering, transmission, and distribution pipeline systems. The GIS data structure is intended to be open source and customized to the pipeline sector. The initiative is backed by a non-profit trade association whose stated mandate is to develop and maintain data, and data interchange standards for the industry. Additional information can be obtained at: http://www.pods.org.

POE: See *probability of exceedance.*

POF: See *probability of failure* or *Pipeline Operator's Forum.*

POFC: Probability of false call. See *detection.*

POI: See *probability of identification.*

Poisson's ratio: An elastic constant of materials, which defines the relationship between the lateral strain and tensile strain due to a tensile stress, in the case of uniaxial loading as shown in the equation:

$$v = \frac{lateral\,strain}{tensile\,strain}$$

where *lateral strain* = strain perpendicular to the applied stress
       tensile strain = strain in the direction of the applied stress

It should be noted that the equation remains the same for compressive stresses. Poisson's ratio is also related to other moduli such as:

$$K = \frac{E}{2(1-2v)}$$

where K = bulk modulus
        E = Young's modulus

Figure 72 shows the relationship between stress and strain expressed by Poisson's ratio.

polarization: The change in potential voltage that results from the presence of current flowing. It is a term that is often used when referring to the effectiveness of an impressed current cathodic protection system. See *off potential.*

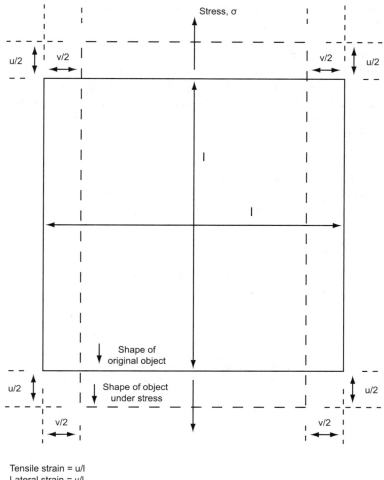

Tensile strain = u/l
Lateral strain = v/l
Poisson's ratio = lateral strain / tensile strain

Figure 72. Various dimensions used for calculating Poisson's ratio. (10)

polarization curve: A curve determined by measuring the potential change as a function of current applied to a metal surface. The potential change that results when the current varies provides a measure of how cathodic and anodic reaction rates are affected by environmental and surface factors (examples include: presence of dissolved oxygen, adsorption, presence of passivating films etc.). Thus, the polarization measurements can be related to the rate of reactions involved in the corrosion process – providing an indication of actual corrosion rates.

polarization decay: The long-term effect seen on the potential once an impressed current cathodic protection system is turned off. Over a period of hours, the measured potential (relative to a reference electrode) will decay. See Figure 1. In the case of buried pipelines, the typical criteria requires that a minimum of a 100 mV decay be observed as one of the indicators of adequate pipeline protection. See also *cathodic protection*.

polarized potential: See *off potential*.

polyethylene tape: A tape coating where the main constituent is polyethylene see *coating* (Table 16).

polly pig: Originally a reference to a specific brand name, it is now a generic term that refers to various soft-bodied urethane-based pigs used for pipeline cleaning. The tools are typically bullet-shaped and can be coated and/or fitted with various accessories, depending on the specific cleaning needs (e.g., wire brushes, hardened metal studs, silicon carbide chips, etc.,).

pooling: The accumulation of liquids in a specific location in a gas pipeline. These locations often coincide with low elevation points, sharp direction changes, and (relatively) sudden increases in diameter (i.e., causing a reduction in gas velocity). Pooling can also occur on uphill sections of a pipeline where the force on the liquid due the forward velocity of the gas equals the backward (downward) force due to gravity. Locations of pooling can be of particular concern in terms of internal corrosion, especially in situations where the composition of the liquid is conducive to increasing corrosion rates. See also *dead leg*.

poor adhesion: A situation where the coating has lost either all adhesion or can be easily peeled from the pipe surface. The defect can be caused by several factors and as a result, good quality control during coating application is the best defense. These key factors include, but are not limited to, surface contamination, incorrect surface profile, incorrect application temperature, and inadequate flow of coating into the surface profile.

poor fit up: See *misalignment*.

porosity: See *weld porosity*.

post weld heat treatment (PWHT): Refers to a range of heat treatments to which welds can be subjected to meet code requirements, relieve residual stresses, reduce hardness, or improve material strength and allow hydrogen diffusion. The key factors to control in heat treatment are rate of heating, final temperature, time at final temperature, and rate of cooling. Of particular note in pipeline applications is the use of heat treatments in cases where hydrogen-induced cracking or stress-corrosion cracking concerns exist; the reduction of residual stresses as well as improvements in material strength can improve the performance of the pipeline with respect to this particular threat.

pot life: The time period that liquid coating is considered to be viable after mixing *but prior to application*. Use of the coating beyond the pot life is likely to result in inadequate protection of the pipe surface. The acceptable period is specific to each coating type and is determined by the manufacturer.

potential impact radius (PIR): In reference to determine high-consequence areas in US regulation, the term refers to the radius of a circle within which the potential failure of a gas pipeline could have significant impact on people or property as determined by

$$r = 0.69\sqrt{PD^2}$$

where r = potential impact radius (ft)
        0.69 = the factor for natural gas
        P = pipeline MAOP (psi)
        D = nominal diameter (in.)

The reader is referred to Section 3.2 of AMSE/ANSI B31.8S for different gas factor constants. (Richer gases have a larger factor.)

poultice corrosion: The localized corrosion associated with debris or other discontinuous build-up on the surface of pipe. Depending on the nature of the deposit (scale, soil etc.,) a highly corrosive environment may exist, especially if moisture is trapped within the area. Also known as deposit attack.

powder coating: A dry factory-applied coating (typically, fusion bond epoxy coating) that does not require a liquid binder. Electrostatic charge is used to attract the particles to the pipe, which is then heat cured to ensure that the coating is bonded to the surface. See *coating*.

PPM: See *parts per million.*

PRCI: See *Pipeline Research Council International.*

precipitation hardening: The strengthening of a metal through the formation of a second phase in the material that is distributed throughout the base material. Specifically, the second phase of the metal is formed by the precipitation from solid solution through specific heat treatment methods. This is the principle mechanism responsible for the strength of ultrahigh-strength steels, aluminum, and titanium alloys.

precipitation inhibitor: A chemical mixture injected into the pipeline to reduce the potential for blockage resulting from solid material settling out of the product stream. Solids that are often a concern are salts and scale, as well as heavier ends of organic compounds (e.g., waxes and asphaltenes). The specific chemistry of the inhibitor varies depending on the pipeline product, flow conditions, and physical geometry that may affect precipitation locations and/or inhibitor injection points.

precipitation strengthening: See *precipitation hardening.*

predictive maintenance: An approach to maintenance planning that relies heavily on monitoring and testing. The extensive use of data is intended to identify the point in time just prior to the degradation of the equipment's performance. Based on this information, servicing is scheduled in an attempt to reduce costs by extending the maintenance period as much as possible.

preliminary in-line inspection report: The initial results of an in-line inspection provided by the vendor identifying the most serious anomalies in a pipeline section. The specifics of reporting times and size/nature of defects identified vary significantly based on the specific contractual agreement, the in-line inspection technology used, the density of anomalies in the pipeline and its length and diameter. However, typical reporting times for (corrosion inspection) preliminary reports are less than one month, and vendors endeavor to identify anomalies that are anticipated to be deeper than 70% or have a burst pressure less than an operator defined threshold (typically as a % of *MAOP*).

PREN: See *pitting resistance equivalent number.*

pre-run questionnaire: A structured review of all available documentation for a given section of pipeline to determine what, if any, specific barriers (see *unpiggable*) to a successful inspection may exist. Table 71 shows typical sources of information.

prescriptive regulation: A philosophical approach to regulating the operations of a pipeline that is very detailed in identifying the nature, frequency and scope of activities to be undertaken. As such, the level of pipeline integrity resulting from this approach can vary significantly depending on the exact condition of the pipeline system; however, the assessment and compliance activities undertaken by the regulator are relatively straightforward. One can argue in this scenario that operators would be motivated to further increase the level of safety of their operations only if there was a financial incentive to do so. This is in contrast to a performance-based philosophy that outlines the required outcomes of pipeline operations (e.g., minimum level of safety), leaving the methods and processes used to achieve the outcomes in the hands of the operator.

pressure: The amount of force per unit area, or stress, the product exerts on the internal surface of the pipe and is a key input parameter for safe design and operation of facilities. Pressure is quoted in a range of units including kilopascal (kPa), bar (bar), atmospheres (atm), and pounds per square inch (psi).

pressure drop: The difference in pressure, typically between two geographically distant points, such as two compressor stations, on a pipeline.

Table 71. Key elements of a pre-(pig)run questionnaire (124)

Source Type	Typical Information
As-built drawings	• Physical restrictions, diameter changes, unbarred off-takes
Previous inspection run report	• Previous experience with debris, operational challenges and pigging procedure • Any tool modifications undertaken to ensure safe passage of tool (e.g., oversized cups to handle large unbarred tees) • Specific considerations regarding previous tracking locations, remote access and run logistics
Previous in-line inspection data	• Confirmation of data from as-built drawings
Repair records	• Identification of any relevant pipeline modifications
Field Personnel	• Identification of any specific operational considerations affecting tool speed, launch/receive and potential for debris • Identification of recent pipeline modifications that could impact success of inspection
Gas control/volume planning	• Scheduling constraints • Inspection tool speed constraints and flow requirements • Specific customer requirements (i.e., service interruption, etc.)

pressure gauge: An instrument used to measure the pressure at a given point. There are two main types of instruments differentiated by the reference pressure that is used: those that measure gauge pressure (in bara) and those that measure absolute pressure (e.g., in barg). For many applications, the pressure as measured relative to atmospheric pressure (i.e., gauge pressure), is most relevant (for example, the output of an air compressor). For other applications, where the measured value is insensitive to changes in atmospheric pressure (e.g., pipeline pressure), the more relevant reading is the absolute pressure (i.e., relative to a vacuum). Thus, the use of the incorrect pressure gauge will results in erroneous readings and process control. A broad range of techniques exist for pressure measurement and guidance on pressure gauge selection can be found through the standard *EN 837-2 : Pressure Gauges: Selection and Installation Recommendations for Pressure Gauges,* published by the British Standards Institute. See *absolute pressure* and *gauge pressure.*

pressure profile: The pressure readings of a pipeline plotted as a function of distance along the pipeline that can be derived from either actual readings or hydraulic simulations. A particularly useful tool when one is attempting to visualize the combined effects of compression/pumping, pressure control equipment, elevation changes, and any other factors that impact how pressure varies in a pipeline system.

pressure recorder: A device used to measure the internal pipeline pressure at a particular location.

pressure reduction: A situation where the operating pressure of a pipeline is decreased in response to a known or suspected integrity threat. In the absence of specific/relevant codes and standards, the size of the pressure reduction is usually based on the assessment of a combination of factors that include, but are not limited to, the nature of the threat, the severity, the certainty of information, and consequences of a disruption/failure.

pressure regulator: A device, usually on a distribution line, that reduces the pressure of gas delivered to a lower pressure acceptable to the customer (i.e.,, customers often do not have the facilities to accept product at the higher pressures encountered on transmission lines).

pressure relief valve (PSV): Valves installed on pipelines, pressure vessels, and other equipment used to reduce the likelihood of an over-pressure situation. That is, the valves are designed to open automatically should a predetermined internal pressure be reached.

pressure surge: A phenomenon that occurs in liquid systems, and at times in gas systems, where a significant pressure wave is created in the pipeline system due to a sudden change in fluid flow or pressure. In the context of pipeline systems this is most often created by the rapid closure of a valve. Because the pressure transient can be significant (i.e., the compressibility of the liquid and elasticity of the pipe become significant in any hydraulic analysis.), care must be taken in determining how quickly valves are cycled to avoid over-pressure of the line.

pressure safety valve (PSV): See *pressure relief valve.*

pressure test: A process where the pipeline is filled with either a fluid (e.g., water or air), sealed, and subjected to a pressure higher than the normal operating pressure. The objective of a pressure test is to either establish an operating pressure limit for a pipeline or detect and eliminate defects in the pipeline. Typically, the pressure is held for several hours to ensure that there are no leaks in the pipeline (this is to be certain that an identified pressure drop is truly the result of a leak and not merely a temperature drop). The procedure is typically used in three scenarios: new pipelines, proving the integrity of existing lines, or proving ability of existing lines for new operating conditions (i.e., increased operating pressure or transportation of new fluid). As with any technique, there are trade-offs in choosing to pressure test a line, as shown in Table 72.

The logistics of hydrostatic testing, such as the choice of test fluid and length of test section(s), are primarily dictated by practicalities. For example, the use of a liquid test fluid such as water is preferable (compared to a gaseous medium) because the energy content of a pneumatic test is many times greater than that of a hydraulic test, and can support very long running fractures in the pipeline if it fails. However, if water is not readily available in remote locations, air may be used. In a similar manner, the length of the test sections may be limited if there are significant elevation changes (particularly if a liquid test medium is being used).

The actual test pressure is a function of the class location, test fluid, and governing codes in the jurisdiction. In cases where B31.8 applies, typical test pressures are defined in Table 73.

Table 72. Advantages and disadvantages of pressure testing (10)

Pros	Cons
• Continuous technique (vs., discrete sections along pipe)	• Potentially destructive method
• Does not require specially designed launcher/receivers or bends	• May require testing in several pipeline sections depending on elevation profile
• Fast results	• Does not locate or assess damage
• Does not require clean line	• Test product may require specific cleaning procedure(s)
• Can be low cost	• Production impact (i.e., outage)
• Low technical requirements	• Only provides a pass/fail type of indication regarding line condition – no additional information regarding remaining defects

Table 73. Standard pressure testing requirements (ASME B31.8) (10)

Class Location	Test Fluid	Test Pressure Min	Max	Maximum Allowable Operating Pressure
Class 1 (Div 1)	Water	1.25 x MOP	None	Test Pressure/1.25
Class 1 (Div 2)	Water	1.1 x MOP	None	Test Pressure/1.1
	Air	1.1 x MOP	1.1 x MOP	Test Pressure/1.1
	Gas	1.1 x MOP	1.1 x MOP	Test Pressure /1.1
Class 2	Water	1.25 x MOP	None	Test Pressure/1.25
	Air	1.25 x MOP	1.25 x MOP	Test Pressure/1.25
Class 3	Water	1.4 x MOP	None	Test Pressure/1.4
Class 4	Water	1.4 x MOP	None	Test Pressure/1.4

where:
- Maximum operating pressure is the highest operating pressure during normal steady-state operation
- Maximum allowable operating pressure is the maximum pressure allowed by B31.8
- Design pressure is the maximum pressure permitted by B31.8 based on materials and location

preventive maintenance: The periodic servicing of pipelines and associated equipment in an attempt to minimize significant downtime and failures to extend its useful life. Various philosophies can be used in establishing the appropriate period and scope of maintenance work; these range from the straightforward application of the manufacturers recommendations to the more sophisticated processes associated with predictive and reliability based methods.

primer: The initial base coat of paint-like material that may be required prior to coating a pipe surface; the purpose of the layer is to ensure that appropriate conditions for bonding of the subsequent coating are in place. While specific needs vary by product and the specific situation, primers are most often required with field applied coatings in recognition of less than optimal conditions for coating application.

prioritization: The ranking of factors or events in specific order of urgency or importance. The *accurate prioritization* is essentially the desired outcome of effective pipeline integrity management programs. The need for increased precision has driven the more recent developments in pipeline integrity risk management (see *risk assessment*).

probabilistic: The group of approaches or methods where uncertainty is explicitly addressed by the use of probability models. Probabilistic methods include *probability of exceedance*, *reliability,* and *risk* analyses.

probability: The most compelling definition of probability is the "degree of belief." When one speaks of the probability of a pipeline failure, the belief regarding the likelihood of an event occurring in a specified future period is expressed. Probability is most often expressed as a decimal less than or equal to one or a percentage less than equal of 100%. Thus, on a basic level, probability is defined as the likelihood, or frequency, of an event, or particular outcome, occurring expressed numerically. That is, an impossible event will have a probability of zero and a certain event will have a probability of 1. In the context of pipeline integrity, the concept of probability is integral to formalized risk management of pipeline integrity (see *risk assessment*). Historical data, usually in the form of summary statistics, often partially establishes the degree of belief about future events. Such data, however, is not the only source of probability estimates, and the process for establishing the probability of a particular outcome can rapidly become quite complex for several reasons. First, the specific likelihood of a particular even occurring may not be known (or known with any degree to certainty) and one may need to establish a range and distribution of probable outcomes (see *probability distribution*). Secondly, added complexity arises from the fact that the range and distribution of likely outcomes may vary further *over time* And/or may be conditional on prior events.

probability density function: See *probability distribution*.

probability distribution: In the case of continuous variables, the probability distribution describes the range of possible values that a random variable can attain and the probability that the value of the random variable is within any (measurable) subset of that range. In the case of discrete variables, the probability distribution identifies the probability of each value of a random variable. Two examples of distribution functions appear in Figures 73 and 74.

In the context of pipeline integrity, establishing probability distributions for likely outcomes is integral to reliability-based methods (see *reliability*) as well as quantitative risk management of pipeline integrity (see *risk assessment*). The optimal choice of PDF for a given variable depends on the nature of the underlying data and requires informed judgment; generally, the behavior of the "tail" becomes especially important as can be seen in Figure 75. The most commonly used probability distributions, and parameters used in risk and reliability analysis, appear in Table 74 (additional detail is available under the entry for type of distribution).

Sample Discrete Distribution

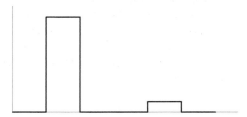

Figure 73. Example of discrete probability distribution.

Continuous (Normal) Distribution

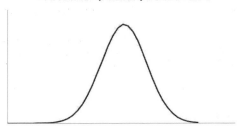

Figure 74. Example of continuous probability distribution.

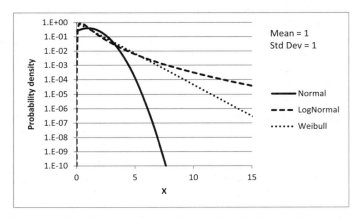

Figure 75. "Tail" behavior for selected probability distributions used in pipeline integrity risk and reliability analysis.

probability distribution function (PDF): See *probability distribution*.

probability of detection (POD): See *detection*.

probability of exceedance (POE): An analytical method that explicitly accounts for uncertainty to maintain the safe operation of a pipeline. Usually, the method accounts for uncertainty only in the depth, length, and burst pressure of a corrosion (or other type of) anomaly. The key parameters of a POE analysis are the depths, lengths, and corrosion rates of corrosion anomalies reported by an in-line inspection run. The uncertainty of depth and length of a corrosion anomaly is estimated either from a comparison of the in-line inspection results to field excavations or from the in-line inspection vendor's stated accuracy. The depth and burst pressure are predicted using the corrosion rate, and a maintenance plan is developed based on the probability that depth exceeds a critical value or that burst pressure falls below a critical value. Figure 76 shows the calculation of POE for depth. The critical value for depth is often 80% of NWT and the critical depth for burst pressure is often MAOP.

Table 74. Common probability functions for pipeline reliability analysis (187)

Type of Distribution	General Shape	Range of Values	Typical/Common Usage
Normal	Continuous distribution with a symmetric bell-shaped curve	$-\infty < x < +\infty$	Widely used to model errors and uncertainty in corrosion depth, pipe diameter, yield strength, tensile strength, and wall thickness.
Beta	Continuous with a positive or negative skewness, define for x between 0 and 1	$0 < x < 1$	Often used to model ratios of success to trials.
Lognormal	Continuous distribution with positive skewness and a "fat" tail, defined for positive x	$0 < x < +\infty$	Has been used to model yield strength of steel, corrosion length, and ratio of maximum depth to average depth.
Exponential	Continuous distribution with a peak at 0 and a negative exponential decaying tail	$0 < x < +\infty$	Has been used to model distribution of corrosion growth rates.
Gamma	Continuous distribution with positive skewness, defined for positive x	$0 < x < +\infty$	Has been used to model distribution of corrosion growth rates when rates vary with time.
Weibull	Continuous distribution with positive skewness tail, defined for positive x	$0 < x < +\infty$	Generally used as an extreme-value distribution. Has been used to model uncertainty in corrosion rates.
Gumbel	Continuous distribution with positive skew, defined all real x		Generally used as an extreme-value distribution. Has been used to model uncertainty in corrosion rates.
Rectangular	Continuous distribution with rectangular shape	$a < x < b$	Often used as a default distribution when only the minimum and maximum values of a parameter are known.

For additional information about the distribution functions described in this table, please see GRI-04/0229, *Guidelines for reliability based design and assessment of onshore natural gas pipelines*, by Maher Nessim and Wenxing Zhou, page D.16.

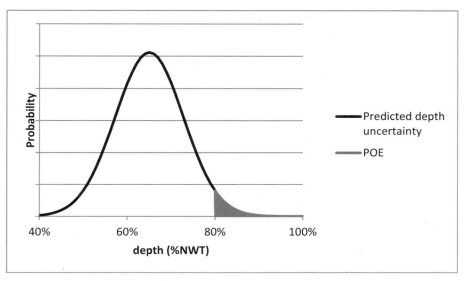

Figure 76. Probability of exceedance is calculated as the area under the probability-density function above the critical depth.

Table 75. Key factors in determining probability of failure (196)

Factor	Description
Exposure	The likelihood of force or failure mechanism reaching the pipe when no mitigation is applied
Mitigation	The actions that keep the force or failure mechanism off the pipe
Resistance	The system's ability to resist a force or failure mechanism applied to the pipe

probability of false call (POFC): See *detection.*

probability of failure (POF): The likelihood that a given pipeline (or pipeline segment) will fail relative to a set of pre-defined criteria. The calculation usually evaluates the chance of one or more failures in a given time period. Table 75 shows three factors that must be assessed to effectively evaluate the probability of failure. The evaluation of these three elements for each pipeline segment results in a probability distribution for failure for that specific segment.

probability of identification (POI): Refers to the likelihood that an in-line inspection system (i.e., combination of tool data collection and processing software) successfully classifies and reports an anomaly's type. For example, POI may be used to quantify the ability of an in-line inspection system to correctly classify internal and external corrosion anomalies.

procedure: A series of steps, tasks and safety precautions taken to accomplish an end goal. In the context of pipeline integrity, this may relate to various different areas including waste handling, welding, pigging, and evacuation/isolation of pipeline product.

product: The commodity that is carried/transported through the pipeline system in question.

program: The grouping of like, or associated, individual projects into a larger cohesive package of work is often termed a *program* for budgeting or reporting purposes. For example, the hydrostatic testing of several pipeline segments may be executed as individual projects; however, the group of all hydrostatic test projects could be termed a hydrostatic test program. Likewise, all of the individual projects undertaken to support pipeline integrity may collectively be termed a pipeline integrity program, in effect, differentiating the work from ongoing operational activity.

protective coating: See *coating.*

PSI: See *pound-force per square inch.*

PSV: See *pressure relief valve.*

public: Relating to, or impacting, the community or the people in the vicinity of the pipeline or associated operations.

public education: The process of undertaking regular contact sharing with property owners, residents living adjacent to the pipeline ROW, and the public at large. The contact can take several forms including, but not limited to individual contact, mailouts, community presentations, and advertising. It is considered to be a significant mitigation strategy for the risk associated with third-party damage to pipelines. It is felt that most third-party damage-related incidents are a result of ignorance of the pipeline in general, its specific location, and potential consequences of damage. Thus, the goal of ongoing contact is to educate key stakeholders along the length of the pipeline and create an awareness that pipeline safety (i.e., avoidance of third party damage) is also in everyone's best interest.

puddle welding: See *weld deposition repair.*

pulse echo: An ultrasonic technique that relies on a single transducer (containing the transmitter and receiver) and the time elapses between the initial signal and reflected signals from both the discontinuity and rear wall. See also *ultrasonic testing.*

pulsed array: See *phased array.*

pump station: The set of facilities on a liquid pipeline that encompasses the pumps and any ancillary equipment used to maintain flow along the pipeline.

pup: A short section of pipe (i.e., less than a typical joint length) used to manage a transition between fittings, change in wall thickness, or a directional change in the pipeline. Often used as a reference point when attempting to locate anomalies identified by in-line inspection tools.

purge: The evacuation of a specific product, or gas, from the pipeline. This is a particularly important operation in situations where flammable products are carried in the pipeline or in situations of confined space entry where sufficient oxygen levels are necessary for worker safety.

PWHT: See *post weld heat treatment.*

Q

QA: See *quality assurance.*

QC: See *quality control.*

qualified: In reference to US regulations for Operator Qualification, qualified means that an individual has been evaluated and can perform assigned covered tasks and can recognize and react to abnormal operating conditions.

qualitative risk analysis: A risk modeling technique where the outcome of the analysis is not comparable to risks external to the pipeline industry (such as vehicle or earthquake facilities). While hard numerical data is used (similar to quantitative risk analysis), the rankings of various inputs and factors used within the analysis are relative to other inputs within the analysis; thus, the results are useful in determining where work needs to be done not whether it needs to be done. See *relative risk value.*

quality assurance (QA): The systems and processes to ensure that quality is maintained. It tends to be a high level system that is proactive and targeted towards preventing defects.

quality control (QC): The activities that are formulated to support the overall quality assurance program. Specifically, the activities tend to be reactive and support the identification of defects that may have been introduced.

quantitative risk analysis: A rigorous mathematical approach applied to the problem of pipeline integrity with the aim of numerically determining the absolute accident frequency and associated consequences. This type of risk analysis is typically the most data-, time-, and therefore cost-intensive to implement. See *risk assessment.*

quenching: The process of holding a metal at a high temperature (i.e., above the temperature at which it would recrystallize – but not melt) and then rapidly cooling it to a predetermined temperature to obtain a very specific grain structure. For example, quenching steel may result in the formation of martensite – a common form of steel often used for tooling.

R

radial sensor: The sensor on an in-line inspection tool that measures the magnetic flux leakage in the radial direction (i.e., the field emanating in direction moving outward from the centerline of the pipe).

radiographic inspection: See *x-ray*.

radius of curvature: The radius of a circle that would (approximately) lie on the curve in question. More specifically, in the context of pipe bends, the radius of curvature is usually indicated relative to the diameter of the pipe. That is, a bend is referred to as "1D" (indicates a bend radius equivalent to one diameter of the pipe), "3D" (indicates a bend radius that is equivalent to three diameters of the pipe) and so forth. The radius of curvature is important in a number of situations – especially when determining whether bend radii in a pipeline are sufficiently large to allow safe passage of inline inspection tools.

random: A characteristic of a number, value, or parameter where the actual value is unpredictable or varies with no set pattern beyond a probability distribution (i.e., it varies based on known, or estimated, probabilities for a range of values). See *probability distribution*.

Rayleigh wave: A specific form of ultrasonic wave that travels along the surface of the material being tested (typically to the depth of one wave length). The particle motion has both transverse and longitudinal components – essentially elliptical in nature. In the context of pipeline applications, Rayleigh waves have been used primarily for the detection and sizing of cracks. See *ultrasonic testing*.

RBDA: See *Reliability-Based Design and Assessment*.

RBI: See *risk-based inspection*.

RCM: See *reliability centered maintenance*.

receiver: The facilities used to extract in-line inspection tools at the end of an inspection run. See *pig trap*.

records management: The systems and processes that are used to identify, store, retrieve and otherwise manage information in a systematic way such that it can be accessed in a timely and accurate manner. In the context of pipeline integrity, records management should encompass documentation that is inclusive of initial construction documents (such as as-built drawings), ongoing monitoring activities (such as cathodic protection records), mitigation activities (such as repairs, sleeve installations etc.), and failure history (time and cause of any leaks and ruptures).

rectifier: A device that converts high-voltage alternating current power (i.e., from a traditional power supply) to lower voltage direct current power. Rectifiers are most commonly used on pipelines in an impressed current cathodic protection systems. Because the device both transforms and rectifies the power, it is often also referred to as a *transformer rectifier*.

reduction: A generic term for a chemical reaction where the molecules of a substance gain electrons. More specifically, reduction is the second part of a corrosion reaction where water molecules gain electrons to form elemental hydrogen as well as generate hydroxide ions (OH^-) – which are then available to form iron oxide to complete the corrosion reaction. See *corrosion*.

reference electrode: A temporary or permanent device to measure the electrical potential at a specific location. In the context of cathodic protection systems, this is the electrode that is used to establish baseline values for the site allowing adjustment for specific conditions that exist; thus the cathodic protection system is designed to achieve a level of protection relative to the potential readings from the reference cell. The reader is referred to *cathodic protection* for a more in-depth treatment of the topic.

reference half-cell: See *half-cell.*

regulator: 1. A government body that has jurisdictional authority over the design, construction, permitting, and operation of a given pipeline system. **2.** See *pressure regulator.*

rehabilitation: The combination of maintenance and improvement activities undertaken to restore a pipeline section to a serviceable condition. This typically is used in reference to pipelines that require extensive servicing, usually, well above and beyond routine maintenance programs.

reinspection frequency: The length of time between in-line inspections of the pipeline. While numerous methods are used in industry to establish the length of this cycle (especially in light of the prevailing maintenance philosophy and the specific condition of each pipeline), there are two main approaches that can be used. One is essentially a fixed interval that is arrived at by analyzing the most severe corrosion/cracking feature and the length of time that would be required for it to grow to failure (based on a conservative growth rate that would be relevant for the pipeline in question). A second approach is essentially risk based and thus establishes the optimal time period between inspections through balancing the probable cost of pipeline repairs vs. the cost of a full in-line inspection. In some jurisdictions a minimum reinspection frequency may be dictated by the governing regulator.

relative risk value: The element of risk (associated with a specific pipeline section, for example) in comparison to other sections where the risk has been calculated using the same methodology. That is, there is no correlation or tie to the absolute value of risk. As a result, this approach is then most useful in prioritizing *where* mitigation activities are to be conducted as opposed to determining *whether* they should be undertaken.

release quantity: The amount of pipeline product released into the environment used in the analysis of consequences of leaks / rupture in the US jurisdiction. Specifically, the term refers to the quantity of spilled material that will trigger an investigation by the Environmental Protection Agency (EPA). The guidelines vary according to the substance (i.e., the more hazardous the substance, the lower the spill limit).

reliability: Defined as $1 - P$, where P is the probability that a component part or equipment associated with a pipeline system will perform satisfactorily under given operating conditions for a specified time period. For pipelines, the probability P is often expressed in terms of a probability of failure per year per kilometer or per year per mile. Variables that may be considered include (but are not limited to): actual operating time, the effectiveness of maintenance, the makeup of the product, the impact of environmental conditions under which the system operates on coatings, CP, compressor performance, and spurious events, such as upsets. Reliability is typically defined in terms of a probability of "satisfactory operation" over a specified time period.

reliability-based design and assessment: An approach to design where the uncertainty in key parameters (such as pipe wall thickness and material strength) is treated explicitly. Rather, key design parameters have statistical variability and are handled using probabilistic methods. The design parameters are selected to meet specific minimum reliability targets – the setting of these targets is often controversial. Reliability is addressed in CSA Z662 and a draft of ASME B31.8S standards. This approach is in contrast to traditional methods where the treatment of uncertainty is implicit – through the use of (often conservative) safety factors.

reliability-centered maintenance (RCM): A risk-based approach to equipment maintenance and capital improvement planning. That is, the overall cost of the equipment operations is optimized by including not only the cost of maintenance but also the probability and consequences of equipment failing. For complex equipment, this approach can be taken on a component-by-component basis to determine the optimal operating costs.

remnant magnetization: The level of magnetization in the pipe wall that remains after the passage of an MFL inspection tool. Remnant magnetization is caused by a hysteresis effect, where the steel "remembers" the past magnetic history. The application of a strong magnetic field to steel causes a magnetization of the steel. When the magnetic field is removed the magnetization of the steel decreases, but does not return to zero. Because the level of magnetization can be significant (but well below saturation levels), variations in material properties (specifically hardness) can often be identified through a second inspection where data are recorded but the permanent magnets have been removed.

remediation: The act of restoring the pipeline to design conditions once a flaw or feature that impacts the integrity of the pipeline is found. Examples include the installation of a pressure containing sleeve to mitigate the risk from extensive general corrosion or the replacement of a short section of pipeline that had been found to be leaking.

remote earth: A reference point for the measurement of cathodic protection potentials. More specifically, the reference electrode is placed at a location sufficiently distant as to remain unaffected by the cathodic protection system itself, or any other cathodically protected facilities in the vicinity. This distance may be on the order of several hundred feet from the pipeline that is being tested.

remote sensing: The broad range of techniques that allow the monitoring of the condition of the pipeline from a distant location (without manual intervention/site visitation). One of the most common examples is the use of satellite imagery to monitor the pipeline ROW for third party incursions/damage.

remotely operated valve: A valve that can be cycled to any position from a centralized control room, i.e., manual operation of the valve is not required.

repair: The act of restoring the pipeline to design conditions once a flaw or feature that impacts the integrity of the pipe line is found. See *repair methods*.

repair criteria: This term is most often used in the context of assessing anomalies identified through in-line inspection. The criteria are most often a set of defect dimensions (and associated failure pressures) which, if exceeded, dictate that the anomaly must be investigated further given the pipeline is likely at significant risk of failure. One of the most common criteria is that any defect with a depth reported to be 70% of the pipe wall thickness requires excavation, further investigation, and potential repair. The term is often used interchangeably with *dig criteria* and (erroneously) with *failure criteria*.

repair method: The technique used to mitigate the risk associated with pipeline defect. The choice of a repair method depends on many factors, including, but not limited to, the characteristics of the defect that is to be repaired, whether the repair is temporary or permanent, whether an outage is possible, the severity of the defect, cost and code requirements. Table 76 shows common repair methods.

Table 76. Common repair methods (6)

Repair Method	Application
Recoat	Simple recoating of the pipeline can be done as a repair for small non-injurious anomalies. The goal of the recoating is to prevent any further damage from corrosion or other environmental damage.
Composite reinforced wrap	For pipelines with blunt-bottomed external defects with depths less than 80% of the wall thickness, composite wraps can be used for to repairs.
Welded sleeve	A metal sleeve can be used to repair a pipeline defect. Pressure-containing sleeves can be used on defects of any depth.
Cut-out	Cut-out repairs are usually used as a last resort, when other repair methods are not applicable. The repair involves the replacement of the damaged length of the pipeline with new pipe.
Depositional welding	Deposition welding is sometimes used to add metal to a damaged area of a pipeline.
Patch	A metal patch over a pipeline defect is sometimes used (though uncommonly, nowadays) as a repair method.

repair procedure: A procedure that itemizes the steps, tasks, and safety precautions taken to ensure that the risk from an identified defect on a pipeline has been mitigated. While each situation is unique and requires a detailed assessment of the specific conditions involved, some of the key elements of a typical procedure would include, but not limited to, requirements for exposing the pipeline (including isolation/depressurizing requirements), materials and techniques to be used for repair, and any QA/QC requirements needed to confirm a successful repair.

reporting requirements: The criteria and standards that apply to the reporting of data resulting from in-line inspections, excavations, and other pipeline-integrity-related services. Often, this data forms the basis for meeting regulatory requirements and is therefore often articulated as a part of the contractual agreement for the services rendered. Table 77 shows some of the key considerations by major services.

reporting threshold: The level at which a testing technique can identify and size defects within the stated confidence limits. For example, vendors with high-resolution MFL tools may not report anomalies that are shallower than 10% of the wall thickness because these types of anomalies cannot be consistently identified and sized.

rerating: 1. The combination of maintenance and improvement activities undertaken on a pipeline section to increase its maximum allowable operating pressure. **2.** The reduction of the maximum allowable operating pressure of a pipeline section based on known factors that have compromised the initial design conditions. (Also known as derating.)

Table 77. Typical reporting requirements for in-line inspection reports and associated excavations (124)

Activity	Key Considerations
In-line inspection	• Accuracy and precision of the size and location of the reported anomalies • Interaction rules • Equation used for burst pressure calculation • Reporting thresholds • Reporting timelines • Tool speed and operations • Factors impacting quality of data collection/analysis
Excavation Data Co llection	• Location of excavation • Length, depth, width of excavation • Location of feature (axial and circumferential) • Nature of feature mapping / record keeping • Documentation of any repair or mitigation work • Reporting thresholds • Reporting timelines

residual stress: The stresses present in the pipe metal that are not associated with, or resulting from, the pressurization of the pipeline system (i.e., they exist in the pipe at ambient, unpressurized conditions). These can be caused by various factors including, but not limited to, welding, fabrication (e.g., cold working of metal), field bends, and mechanical damage. Residual stresses can play a critical role in both the prevention and exacerbation of pipeline-integrity-related issues. For example, the introduction of residual surface stresses through shot peening can provide a protective effect against stress-corrosion cracking; however, residual stresses from a field bend can reduce the ability of the pipe metal to deal with ground movement in the area of cold work.

resistivity: Generally, resistivity refers to the resistance to current flow of a material or soil. In the context of pipeline integrity, the term most often refers to the soil resistivity. Soil resistivity is one of the key environmental conditions affecting external corrosion. High resistivity soils are usually not corrosive, and low resistivity soils tend to be highly corrosive (see Table 78). Soil resistivity is also one of the key considerations for the design and installation of cathodic protection systems.

resolution (in-line inspection): The smallest size increment of length, width, and depth of anomaly that a smart tool is able to detect.

Table 78. Corrosivity dependence on soil resistivity (18)

Resistivity (ohm-m)	Corrosivity
> 50	Mild
20 - 50	Moderate
7 - 20	Corrosive
< 7	Very corrosive

rest potential: See *corrosion potential.*

Reynolds number: A dimensionless basic parameter used in fluid flow calculations to provide an indication of the nature of the product flow (i.e., laminar vs., turbulent). Lower values are indicate of flows that are more *laminar* in nature whereas higher Reynolds number values are indicative of more *turbulent flow*. The equation is essentially a ratio of inertial forces vs., viscous forces and for a pipe is defined as follows:

$$Re_D = \frac{4q_m}{\pi \mu D}$$

where Re_D = Reynolds number
             q_m = mass flow rate
             π = universal constant (3.14....)
             μ = absolute viscosity of fluid
             D = internal pipe diameter

rich gas: Pipeline product that is composed primarily of methane, but also contains significant amounts of higher (heavier) hydrocarbons. The primary concern from a pipeline integrity perspective is ensuring sufficient toughness of the pipeline steel for richer gases. This requirement stems from the decompression behavior of rich gases - which is much slower relative to methane – providing more energy/driving force at a crack tip during a failure. This effect is exacerbated when there is two-phase product flow (both liquid and gas). Because heavier hydrocarbons become liquid at lower pressures, two-phase flow is more likely in pipelines with rich gas.

right-of-way: A pipeline is installed within a strip of land referred to as a right-of-way. The pipeline company, through lease or purchase, acquires rights to use this land for the construction, operation, and maintenance of its pipelines; however, ownership of the land remains with the landowner (if leased).

risk: A measure of the safety of a pipeline (or other structure), which considers the probability of failure and the consequence of that event. For quantitative risk, risk (R) is defined as the product of the probability of occurrence (P) and the consequence (C).

$R = PC$

Note that for quantitative risk, both the probability and the consequence must be a measureable quantity.

Qualitative risk is often used for applications where the probability of occurrence or the consequence cannot be easily quantified. Table 79 shows a risk matrix for assessing risk.

Table 79. Qualitative risk matrix (31)

		Increasing probability of occurrence			
		Low			High
Increasing consequence	Low	Low risk	Low risk	Moderate risk	Moderate risk
		Low risk	Moderate risk	Moderate risk	Moderate risk
		Moderate risk	Moderate risk	Moderate risk	High risk
	High	Moderate risk	Moderate risk	High risk	High risk

risk assessment: A systematic process in which potential hazards from facility operation are identified, and the likelihood and consequences of potential adverse events are estimated. Risk assessments can have varying scopes and be performed at varying level of detail depending on the operator's objectives. One of the more important distinctions is whether the risk assessment is qualitative or quantitative. (See *qualitative risk analysis* and *quantitative risk analysis.*)

risk management: An overall program consisting of identifying potential threats to an area or equipment; assessing the risk associated with those threats in terms of incident likelihood and consequences; mitigating risk by reducing the likelihood, the consequences, or both; and measuring the risk reduction results achieved.

risk mitigation: The group of activities and processes used to limit either the probability or the consequence of a specific threat. For example, the installation of a pipeline coating is a risk mitigation mechanism for various forms of cracking and corrosion because it is known to reduce, and in some cases stall, corrosion growth, whereas purchasing insurance mitigates the consequence of the resulting pipeline failure.

risk modeling: See *risk assessment.*

risk reduction: See *risk mitigation.*

risk tolerance: The acceptable level of exposure to a given consequence of an event (or set of consequences) that an operator is willing to accept. The term is most often used in the context of absolute risk. Specifically, the level of risk that an operator is willing to accept – implying that any hazard resulting in a higher level of risk would be subject to some mitigating action to bring the risk level down to acceptable levels.

river bed erosion: The gradual wearing away of the material (sediment or rock) at the bottom of a river. Sufficient erosion can pose an integrity threat in that it can result in exposure of the pipe and a potential loss of material (and therefore support) under the pipe.

river bottom profile: 1. The profile of the channel through which a river flows. The profile, combined with water velocity, is a critical factor in determining the rate and nature of erosion in the area. **2.** The profile of maximum corrosion depth versus axial distance through an area of corrosion; generally chosen to represent the most critical (i.e., deepest) path in the area (see Figure 77). The assessment of a river bottom profile for generally corroded areas is a critical step for calculating burst pressure values using and iterative effective area approach such as *RStreng®*.

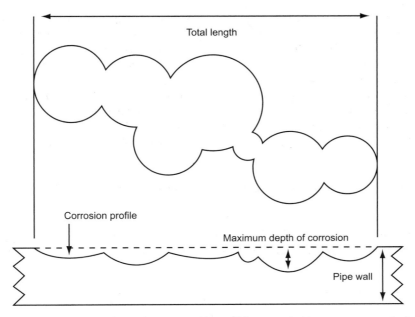

Figure 77. Example of river bottom profile used for corrosion burst pressure calculations (6).

river scour: The overall erosion of the river bed by secondary currents that result from the combination of high velocities and bends or the confluence of channels. Typically, localized scour is of a greater concern than overall erosion of the river bed as localized increases in water velocity significantly increase the ability of the water course to erode material from its bed. The key concern associated with river scour is that extensive scour can expose the pipeline leading to increased integrity related concerns such as mechanical damage and unsupported free spans.

river weights: A concrete weight used for buoyancy control at river crossings and in areas where a significant amount of water remains in the ditch during construction (i.e., it cannot be practically be removed). The ditch is excavated to be sufficiently wide and deep to accommodate this style of weight. The weight typically has a felt coating on its internal surface to minimize damage to the pipe (and coating). The two halves of the weight are applied to the line and bolted on prior to the pipe being laid in the ditch.

robotic tools: Inspection tools that are fundamentally designed to be self-sufficient in terms of locomotion and data collection. Specifically, the tools do not rely on a tether or flow within the pipeline section to operate effectively. While the actual inspection technique will vary (i.e., ultrasonic, MFL, etc.,) these tools are typically best suited to applications where pipe diameter and pipe length are relatively small. Another factor that often leads to the selection of a robotic tool is pipe geometry (i.e., the number/nature of bends in a pipeline section) where the use of a tethered tool is impractical.

rock shield: Material that is used directly over a pipeline section to prevent damage from rocks that may be present either in the ditch or backfill material. Several commercial products are available; they are typically made from some type of PVC.

Rockwell hardness: One of several tests used to characterize material properties where a material's resistance to penetration by an indenter is measured. There are variations of the test, which are indicated by a single letter following the test name (i.e., Rockwell "A," Rockwell "B," etc.). Compared to other hardness tests, deformations in this test are relatively large (i.e., 8-10% strain). It is possible to relate test results to material strength for general classes of materials (quantitative correlation of hardness values to material strength across all classes of materials is not practical). The test does have the advantage of being relatively straightforward and well understood for specific material classes, such as carbon steel (2). See *hardness* for a comparison of major test methods.

root bead: A weld bead that extends into or includes part or all of the joint root.

root pass: 1. The first weld that is laid down during a welding operations (Figure 78). 2. A weld pass made to produce a root bead. See *weld passes*.

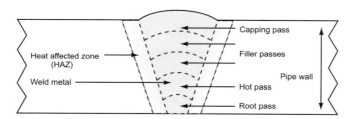

Figure 78. Root pass on a weld.

root undercut: An undercut defect that is introduced adjacent to the root or pass during the welding process. See *undercut*.

ROW: See *right-of-way*.

ROW access: **1.** Permission to move across private land to reach the pipeline during the course of normal operations, routine, or extraordinary maintenance. **2.** The path, or roadway, used to reach the pipeline during the course of normal operations, routine or extraordinary maintenance.

ROW crossing: The point at which a utility (e.g., pipelines, power lines etc.,) or other significant infrastructure (e.g., roads) intersects a pipeline's route. Depending on the nature of the crossing, extra precautionary measures may be required. For example, power lines crossing the ROW may disrupt cathodic protection, or major roads may require the installation of heavy-walled pipe.

RPR: See *rupture pressure ratio.*

RStreng®: A proprietary computer program to calculate the remaining strength (hence the name) or burst pressure of a pipeline. The Conservatism of B31G and widespread use of PCs led to development of the RStreng® Software. The software is proprietary to *PRCI*. However, the equations on which RStreng® is based, are not proprietary and a free download (spreadsheet) is available from Kiefner and Associates, Inc.'s Pipe Assessment (KAPA) webpage (www.kiefner.com/KAPA2006.xls)

The RStreng® equation is based upon the same effective area approach as B31G and requires detailed defect geometry input. This allows it to automatically address complex geometry issues. See Figure 79. Table 80 shows strengths and weaknesses of using this method. Also see *B31G.*

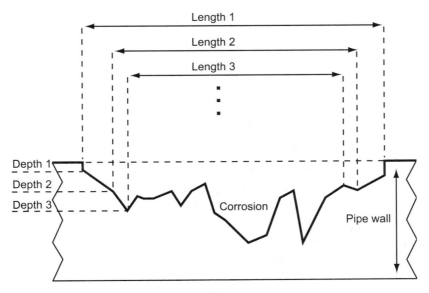

Figure 79. RStreng® input requirements (6).

Table 80. Advantages and disadvantages of RStreng® technique (8)

Advantages	Disadvantages
• Significantly less conservative than B31G and will result in fewer cut outs and repairs • Validated against the experimental database • Automatically accommodates complex geometries	• Requires detailed defect profiles • Does not address axial pipe loads • Has scatter in predictions and is not always conservative in predicting failure

rubbing: A representation of the corroded external surface of a pipeline made by placing paper, or Mylar over the surface and rubbing the paper gently with a marking agent, such as charcoal. While this technique does not provide detailed information regarding the depth of the flaw(s), it provides a good visual indication of the shape and general topography.

rupture: The instantaneous tearing, that is a *running fracture*, of pipe material causing large-scale product loss and immediately impairing the operation of the pipeline. Specifically, the tear in the pipe steel will continue to grow until the depressurization front moves past the growth point of the defect. The tear will stop once there is no longer enough energy left in the pressurized product to support continued crack growth. In the case of large-diameter gas lines, rupture is often associated with a fire as released product is consumed. It should be noted that usage is not consistent and at times the term is used interchangeably with *burst*; the other mode of failure is *leak*.

rupture pressure ratio (RPR): A ratio of the predicted burst pressure calculated by an analysis criterion (e.g., ASME B31G) to the pipeline pressure at specified minimum yield strength. The ratio is typically used as a method for identifying and prioritizing defects that may require repair. For the modified B31G burst pressure equation, RPR is calculated by

$$RPR = \frac{\sigma_f}{\sigma_{smys}} \left(\frac{1-(.85)(d/T)}{1-(.85)\left(\dfrac{d}{TM}\right)} \right)$$

where
σ_f = flow stress
σ_{smys} = specified minimum yield strength
d = depth of corrosion
T = wall thickness of the pipe
M = *Folias factor* to account for bulging

rural area: An area outside the limits of any incorporated or unincorporated city, town, village, or any other designated residential or commercial area such as subdivision, a business, or shopping center or community development. Rural areas typically are Class 1 or 2 locations.

rust: The red/brownish substance remaining after the oxidation of carbon steel or iron in the presence of oxygen and water, fundamentally indicating the breakdown of the base material.

S

sacrificial anode: This category of anode used in cathodic protection systems relies on the fact that the anode material oxides more easily than the metal (or pipe) that is to be protected. Thus, the anode must oxidize almost completely before the pipe, the less active metal, will be subject to corrosion processes. Zinc or magnesium alloys, packed in low-resistance bentonite fill, are the most commonly used materials. This type of cathodic protection system is best suited for congested areas where the low current output reduces the potential for interference problems.

sacrificial system: See *galvanic (cathodic protection) system.*

saddle weights: A concrete weight used for buoyancy control in areas where conventional construction and pipe-laying methods can be used (i.e., no significant water is present in the ditch during construction and the ditch can excavated to sufficient width and depth to accommodate this style of weight). The weight typically has a felt coating on its internal surface to minimize damage to the pipe (and coating) and is typically placed directly onto the line with no direct attachment to the pipeline.

safety factor: A factor incorporated into pipeline design calculations to ensure that the resulting design is conservative to ensure safe ongoing operations. For pipelines the safety factor is less than one and is the factor by which the maximum operating pressure must be reduced below the yield point of the pipe. Safety factors for pipelines are divided into four separate factors: the design factor (F), the location factor (L), the joint factor (J) and the temperature factor (T). The product of the factors, $FLJT$ give the total safety factor. The values of the factors depend on population density in the vicinity of the pipeline, the type of welding, and the operating temperature of the pipeline.

sag bend: The concave section of a bend on the pipeline that accommodates a change in elevation (as opposed to a change in the horizontal plane).

sagging: Refers to the runs that can be seen on the pipe surface if poor technique (or incorrect metering) is used during the application of liquid coating. Specifically, if incorrect metering is used or if the coating is applied in layers that are too thick, runs and sagging can be observed.

sandblast: The process of subjecting the pipe surface with sand traveling at high velocities. The primary application in hydrocarbon pipeline systems is for pipeline surface preparation prior to coating and/or inspection.

satellite imaging: The photographic documentation of the ROW using satellites. The process can be used for monitoring ROW incursions and potential third party damage on the pipeline.

saturation: See *magnetic saturation.*

SAW: Refers to submerged arc welding. See *weld.*

SCADA: See *supervisory control and data acquisition*

scale: Hard deposit on the internal surface of the pipeline formed as a result of the combination of the product, operating temperature, and pressure. Scales can be formed from various substances, but are more commonly composed of calcium carbonate or calcium sulfate and tend to be associated with pipelines that are close to production formations where the ability to refine the product is limited (e.g., offshore pipelines). Pipeline scale is of particular concern because it effectively reduces the internal pipeline diameter decreasing flow capacity. Further, it is very difficult to remove and poses a challenge to internal inspections for regular maintenance.

scaling pig: A cleaning tool specifically designed to remove scale build-up on the internal pipe wall. Typically, these tools will have metallic brushes or spikes to ensure that the hard scale can be effectively dislodged.

SCC: See *stress corrosion cracking.*

SCC density: A reference to the magnitude of circumferential spacing of SCC cracks within a colony. Table 81 shows the definition as set forth in the SCC Recommended Practice published by *CEPA.***SCC – near-neutral:** See *transgranular SCC.*

SCC – high pH: See *intergranular SCC.*

schedule: See *pipe schedule*

Table 81. SCC Density definitions (24)

SCC Density	Approximate Circumferential Spacing
Sparse	< 0.2 x wall thickness
Dense	< 0.2 x wall thickness

scour: See *river scour.*

scraper pig: See *cleaning pig.*

scraper tool: See *cleaning pig.*

seabed scouring: The overall erosion of the seabed by water currents (and ice in polar regions) that result in relatively abrupt changes in seabed topography. Typically, localized scour is of a greater concern than overall erosion of the seabed as it can causes localized increases in water velocity and can significantly increase the rate of erosion. The key concern associated with seabed scouring is that extensive scour can undercut submerged pipelines leading to free-span sagging and related concerns such as mechanical damage. Also caused by ice in polar regions.

seam misplaced trim: A situation where misalignment of the weld trim grinding tool results in an axial groove or depression along one side of the seam weld of ERW pipe. This type of feature is generally not repaired because it does not typically pose a structural concern.

seam over trim: A situation where excess grinding of the weld flash results in an axial groove or depression on the center line of the seam weld of ERW pipe. This type of feature is typically insignificant from a structural point of view and does not normally require repair.

seam under trim: A situation where insufficient grinding of the weld flash results in a raised linear feature on the center line of the seam weld of ERW pipe. This type of feature is typically insignificant from a structural point of view and does not normally require repair.

seam weld: A longitudinal weld on a pipeline section that is a result of the fabrication process (i.e., specifically the join that is required when metal plate is rolled into a cylindrical shape to form the pipe join).

seamless pipe: Pipe that is manufactured by pulling a solid cylindrical billet over a mandrel to create the pipe; with this technique, a seam weld is not required. Seamless pipeline can pose a challenge to MFL inspections because the fabrication process can cause the inspection signal to contain a high level of *noise*. Results of seamless pipeline inspection may have a lower *probability of detection*, a higher *probability of false call*, and reduced *sizing accuracy*.

segment: See *pipe segment.*

self-propelled: See *robotic tool.*

sender: See *launcher.*

sensor: The mechanical portion of an inspection tool that contains the hardware used to take measurements for detecting anomalies in the pipeline. The term is not specific to an inspection technique, but is most commonly used in the context of in-line inspection tool mounted ultrasonic or MFL technologies. Figure 60 shows a typical high-resolution MFL smart tool, representing the full magnetic circuit with the sensor head specifically identified.

separation pig: Typically a bi-direction tool used to separate differing fluids during pipeline batching operations or hydrostatic testing.

service line: A distribution line that transports product from a common source of supply to an individual residential/small commercial customer, or to multiple residential or small commercial customers served through a meter header or manifold. A service line ends at the outlet of the customer meter or at the connection to a customer's piping, whichever is further downstream, or at the connection to customer piping if there is no meter.

shall: The term used in most industry standards to indicate those practices that are considered mandatory (versus those that are recommended or considered best practice but not considered to be a requirement). See also: *should.*

shallow groundbed: Series of cathodic protection anodes installed at a depth of 50 ft (15m) or less.

shear modulus: An elastic constant of materials that defines the relationship between the shear stress and shear strain. Figure 80 shows the relationships between stress and stain expressed by the shear modulus, which is represented as:

$$G = \frac{\tau}{\gamma}$$

where τ = applied shear stress
γ = shear strain (tan θ; where θ is the angle of the deformation)

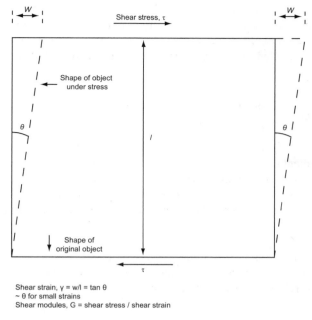

Shear strain, γ = w/l = tan θ
~ θ for small strains
Shear modules, G = shear stress / shear strain

Figure 80. Input values for calculation of shear modulus. (10)

shear waves: Also called transverse waves, they are a form of elastic body wave used in ultrasonic testing. Shear waves are characterized by the motion of atoms and molecules in the bulk material moving in a direction *perpendicular* to the direction of the wave propagation. Shear waves do not travel in liquids or gases, a limitation that poses a challenge in transmitting the shear-wave energy into a pipe wall. See *ultrasonic testing* for additional detail or reference for a more in-depth discussion of the topic.

shielded metal arc weld (SMAW): See *weld.*

shielding: A situation where disbonded coating prevents cathodic protection from reaching the pipe surface, *shielding* the pipe material from the beneficial effect of reducing or eliminating corrosion rates. It should be noted that not all coatings interfere with cathodic protection systems, and the shielding effect is a concern primarily with polyethylene tape coatings. This term is often used interchangeably with the terms *cathodic shielding* and *electrical shielding*.

shop coat: Coating that is applied at a fabrication facility allowing for optimal quality control (as opposed to the field application of coating where quality control is more difficult). This term is used interchangeably with mill-applied and factory-applied coating.

shore hardness: A standardized non-destructive test used to determine the hardness of a material. Hardness is a measure of a material's resistance to plastic deformation or penetration. Shore hardness relies on a durometer and is usually the method used for highly elastic materials (such as rubbers and elastomers). It is an important parameter to control and assess on pipeline coatings specifically because harder materials are often more susceptible to various forms of cracking. The hardness test estimates hardness by pushing the surface of a material with a diamond indenter of a specific size and shape and with a specific force. The hardness is estimated by examining the size of the indentation and can be related to *ultimate strength* through an empirically determined equation.

shot blasting: See *shot peening*.

shot peening: A surface treatment that is used to relieve residual tensile stresses on the external pipe wall. During the process, small spheres (either metal or ceramic) are "shot" towards the pipe surface with sufficient force to introduce residual compressive stresses on the surface of the material. Benefits of the method included improved fatigue life.

should: The term used in industry standards to indicate those practices that are preferred, but for which operators may determine that alternative practices are equally or more effective. The term also refers to practices where engineering judgement is required. See also: *shall*.

shrink sleeve: A non-pressure-containing polymer sleeve, typically used at girth welds, as a pipe surface coating. Sleeves are typically made of thermoplastic polymer and rely on the use of heat to ensure a good bond to the pipe surface.

side bend: A pipe bend that facilitates a change in direction in the horizontal plane (as opposed to the vertical direction).

signage: Posted warnings indicating or highlighting a presence of a structure and/or hazard. The most relevant reference in the context of pipeline integrity is signage indicating the presence of a pipeline in a ROW in an attempt to reduce the potential for third party damage.

significant SCC: The NACE RP0204-2004 criterion defines "significant" size SCC as one for which the deepest single crack in the colony exceeds 10% of nominal wall thickness and, assuming the "interacting length" is equivalent to a single crack, is 75% of the critical length for a 50% through-wall crack operating at a pressure that generates a stress of 110% SMYS perpendicular to the crack. This definition was previously also supported by CEPA but has since been revised into a four category classification system that appears under *stress corrosion cracking*.

sizing accuracy: Refers to the statistical tolerances associated with the particular technique in establishing the dimensions of any identified anomalies. The specification of sizing accuracy requires a tolerance and a confidence level. For example, most high-resolution magnetic flux leakage tools have a sizing accuracy of +/- 10% 80% of the time, when establishing the depth of pipeline anomalies. In this example "10%" is the tolerance and "80%" is the confidence level.

sizing plate: A metal disc mounted on a mandrel or cleaning pig to determine if there are any obstructions in the pipeline that would interfere with the passage of an in-line inspection tool. The size of the plate is typically the minimum diameter that the inspection tool can pass through without damage. After the passage of the pig, if the plate is undamaged, then the diameter of the pipeline has a sufficient diameter to allow the passage of the ILI tool. If a plate is damaged, the nature of the damage is used to determine how to proceed. Minor damage may be considered "acceptable," whereas more significant damage may require that a more sophisticated caliper tool is run prior to the smart tool.

slag inclusion: 1. The incorporation of non-metallic material into the pipe wall through the use of contaminated ingots for production. Typically, these inclusions are not structurally significant; however, the exact nature and size of the inclusion will ultimately govern any threat to pipeline integrity. Surface-breaking inclusions are more of a concern in terms of creating coating holidays and/or corrosion acting as initiation points for corrosion. Slag is a by-product of the smelting process and is typically composed of metal oxides and sulfides. **2.** The incorporation of non-metallic material into the weld area through improper welding technique. Typically, these inclusions are not structurally significant; however, the exact nature and size of the inclusion will ultimately govern any threat to pipeline integrity.

sleeve: Also called full encirclement sleeve, this is one of the most widely used methods of general repair for defects in on-shore pipelines. Sleeves are effective because they restrain bulging and results of research shows that sleeve repair can restore pressure capacity to 100% SMYS in a cost-effective manner. There are two main types of sleeves; Table 82 shows their key characteristics and considerations.

sliver: Thin elongated anomalies caused when a piece of metal is rolled into the surface of the pipe. A sliver is usually attached at only one end. In MFL inspections, a sliver is sometimes called a lamination. Small superficial anomalies are usually not a structural concern; however, they can cause holidays in thin film coatings. Slivers can be removed by grinding.

slope creep: A relatively commonly encountered geotechnical hazard where there is a gradual movement of the soil mass in the down-slope direction. The phenomenon can be encountered on both frozen and unfrozen slopes and results in a gradual loading and eventual deformation of the pipe subject to the movement.

slope grading: A common technique used to stabilize locations where the gradient is such that ground movement is likely to occur. Specifically, the technique involves establishing a stable geometry by flattening portions, or all, of the slope (thereby reducing the amount of ground force driving the movement). Where space permits, this technique is commonly used in conjunction with a stabilizing toe (earthen) berm to provide additional support against downward movement of slope material.

Table 82. Comparison of common pipeline repair sleeve types (6)

Characteristic	Type A	Type B
Pressure Containment	• No (sleeve ends not welded) • Reinforcement only	• Yes (sleeve ends welded)
Fabrication	• Rolled plate • Fabricated from pipe • Single-V butt welds - requires just more than half pipe circumference • Fillet welded overlapping side strips - can accommodate gaps	• Rolled from plate • Fabricated from pipe • Single-V butt welds - requires just more than half pipe circumference • Fillet welded overlapping side strips not recommended for Type B sleeves • Must ensure acceptable chemical composition of sleeve (i.e., carbon equivalent to reduce risk of hydrogen cracking)
Principle of Operation	• For relatively short defects [$L < \sqrt{80Dt}$], effectively functions without being a high-integrity structural member • Must restrain bulging therefore no gap between sleeve and pipe permitted; methods for achieving this include: • Pressure reduction • External loading (force fit) • Use of fillers that harden	• Restrains bulging of the defective area • For leaking defects or for defects that will eventually leak (i.e., internal corrosion), contains pressure • For relatively long defects [$L>\sqrt{80Dt2}$], effectively functions as a high-integrity structural member
Key Design Considerations	• For relatively short defects [$L <\sqrt{80Dt2}$], sleeve can be thinner than pipe but, $t_{sleeve} > 2/3\, t$ • For long defects [$L > (20Dt)1/2$], sleeve should be as thick as pipe sleeve	• Sleeve should be as thick as pipe (or thinner but stronger) • Sleeve should extend at least 2 in. (50 mm) beyond ends of defect • Annular space must be pressurized to relieve stress in pipe wall
Advantages	• Attractive for repairing longitudinally oriented crack-like defects as the driving force for crack growth is removed	• Suitable for circumferentially-oriented defects • Can be used for leaking defects or for defects that will eventually leak (e.g., internal corrosion) • Can be used for cracks • Annular space between sleeve and pipe protected from corrosion
Disadvantages	• Not for circumferentially-oriented defects • Not for leaking defects or for defects that will eventually leak (e.g., internal corrosion) • Annular space between sleeve and pipe difficult to protect from corrosion (for this reason, some companies weld the ends of "Type A" sleeves)	• More costly and complex repair and installation

slope inclinometers: A ground-based technique, also referred to as slope gauges, used to measure the movement of a slope over time. The technique involves the installation of a custom-built PVC tube installed through a borehole; an electronic probe is then raised through the tube, with readings taken at small increments, to determine the deformation of the tube. This data can then be used as an indicator of the overall slope movement. The technique is well established but is prone to wildlife interference, has an environmental impact, and requires a field presence to take measurements. New installations may be required frequently where relatively high rates of movement are occurring.

slug flow: A variant of *wavy stratified flow;* that is, a form of *multiphase* flow where a liquid and gas phase form separate layers as they flow along the length of the pipeline. However, the nature of this *flow regime* is such that the gas phase causes sufficiently large waves in the liquid phase to create short sections of single phase (i.e., liquid) flow. Figure 81 shows this flow regime.

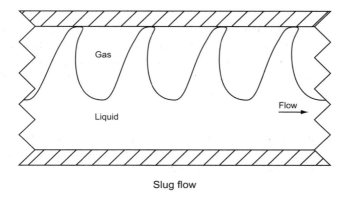

Slug flow

Figure 81. Slug flow. (10)

smart pig: A class of internal inspection tool that has some capability for detecting, sizing, and locating pipeline anomalies using non-destructive techniques and onboard electronics (as compared to a cleaning or sizing tools, which only indicate potential pipeline anomalies through tool damage and do not have an ability to identify the location of potential pipeline defects). See *pig* for a more detailed discussion of smart pigs.

smart tools: See *pig.*

SMAW: Refers to shielded metal arc welding. See *weld.*

SMLS: See *seamless.*

SMYS: See *specified minimum yield strength.*

societal risk: Usually used in the context of quantitative risk and reliability-based methods for establishing maintenance programs. As an example, Canadian Code CSA Z662 (Annex O) defines societal risk as "A measure of the risk where the consequence considered is a function of the expected number of fatalities occurring due to pipeline failures." This is in contrast to *individual risk.*

Society for Protective Coatings (SSPC): Originally known as "The Steel Structures Painting Council" when it was established in 1950 The society is a non-profit association concerned with all aspects of "….the protection and preservation of concrete, steel and other industrial and marine structures and surfaces through the use of high-performance protective, marine and industrial coatings." .The society adopted its current name in 1997 to more accurately reflect the nature of coatings technology. Over the course of 60 years, SSPC has grown to include ~9,000 individual members, 30 organizational members and numerous local chapters internationally – across a range of industries. (189)

SOHIC: See *stress oriented hydrogen induced cracking.*

soil classification: Used to define and describe the nature of the soil found – typically along the pipeline right-of-way or in the ditch during pipeline excavation. The term is often used interchangeably with *soil type* and *soil texture.* No standard classification system exists, but several sources provide guidance in the form of recommended practices such as the one published by CEPA in 2007.

soil creep: The gradual movement of large sections of soil material that unexpectedly stress and stain the pipeline. Most often this situation occurs on slopes and is often indicated by the presence of cracks or slumps on the surface.

soil drainage: A key parameter in defining soil types because it indicates the soil's permeability to water. Commonly used classifications for soil drainage in the pipeline industry are imperfectly, poorly, very poorly, moderately well, and well drained. The identification of soil types can then be used as one of the inputs in determining areas susceptible to various integrity threats such as *corrosion* and *stress corrosion cracking* – particularly as part of a *direct assessment* methodology.

soil profile: A two-dimension cutaway section (also referred to as a terrain/ground profile) perpendicular to the surface of the earth allowing a detailing/documentation of the nature of soil at various depths on the right-of-way or in the pipe ditch. Key information, especially in the context of pipeline integrity, includes the nature of drainage, resistivity, pH, etc., and how these characteristics vary with depth.

soil resistivity: See *resistivity.*

soil texture: See *soil classification.*

soil-to-air interface: The point (also known as ground/air interface) at which a buried pipeline enters or exits the ground. This interface can be of particular importance for pipeline integrity because it represents an area on the pipeline that is often subject to a wide range of conditions (e.g., moisture, temperature, and soil stresses) that can significantly reduce the long-term performance of coatings, cathodic protection, and the pipe wall itself.

soil type: See soil classification.

soils model: A tabulation of key soil properties encountered along the pipeline right-of-way (ROW) – historically used as an aid to predict pipeline locations potentially susceptible to SCC. The tabulation would entail an assessment of the following soil properties: drainage/moisture levels, pH, soil type, topography, texture, and resistivity. This term is often used interchangeably with terrain model.

solid cast pig: A type of utility pig. See pig for a detailed discussion of this class of tools.

solidification cracking: The cracking that results as liquid metal cools and solidifies. It is most relevant to the welding process where an internal crack is introduced due to a high depth-to-width weld-bead ratio or as the result of impurities being present in the weld pool. Any other situation that produces excessive constraint/stress on the weld area during the solidification process can produce this type of defect.

solvent cleaning: Cleaning operations, using chemical compounds, that are undertaken to remove solids in the pipeline. The method is typically used in situations where the material is not readily removed through mechanical cleaning methods. Specifically a substance with a high affinity for dissolving the deposited material (such as waxes, asphalts, or tars) is used in a batching process to clean the pipeline. The choice of solvent used is usually specified based on line conditions; however, generally, the substance chosen should be dispersible in the carrier, penetrate deposits that coat the pipeline, suspend solids, carry the deposits out of line, and settle and separate for reuse and/ or disposal. Typical substances used in pipeline applications include aromatics (such as xylene, toluene, or condensates) and alcohols (other substances, such as acids, are typically not practical for pipelines due to the long soak time that is required).

sour corrosion: The broad range of corrosion types that occurs in an environment where significant levels of H_2S exist in the product. See corrosion.

sour crude: Crude oil that contains more than 1% total hydrogen sulfide (H_2S). There are implications for pipeline design from an integrity point of view. As such, both safety and materials considerations are significantly greater for a pipeline carrying sour product (vs. one carrying sweet product). See hydrogen sulfide.

sour gas: Natural gas that contains measurable amounts of hydrogen sulfide (H_2S). There are implications for pipeline design from an integrity point of view. As such, both safety and materials considerations are significantly greater for a pipeline carrying sour product (vs. one carrying sweet product). See hydrogen sulfide.

sour service: The term refers to any pipeline used to move hydrocarbons that contain sufficient quantities of hydrogen sulfide to warrant the consideration of additional safety measures and specialized materials (see *sour gas* and/or *sour crude*).

spalling: Refers to the flaking of pipeline coating that could result from several mechanisms, including poor application and pipeline deformation. The standard coating test associated for this aspect of coating performance is *ASTM G70 - 07 Standard Test Method for Ring Bendability of Pipeline Coatings (Squeeze Test)*.

specified minimum yield strength (SMYS): The lowest yield strength of the prescribed specification under which the pipe is purchased from the manufacturer. See *stress-strain curve* for detailed discussion regarding key material properties and specification for pipeline steels.

speed control: The physical mechanism that allows the effective in-line inspection of (gas) pipelines that have high product velocities. The mechanism enables an inspection tool to travel down the pipeline at an optimal or near optimal speed by allowing and controlling the amount of product bypassing the tool.

speed excursion: See *velocity excursion*.

spherical pig: A type of utility pig. See *pig* for a detailed discussion of this class of tools.

spiral welded pipe: Pipe that is manufactured by continuously forming strips of steel plating such that a section of pipe has a single, helical seam. The presence of spiral welded pipe in a pipeline can be significant from a pipeline integrity perspective when using magnetic flux leakage (MFL) inspection tools, because spiral welded pipe tends to be magnetically "noisy," making defect detection and sizing more difficult.

spool: A section of pre-fabricated pipe, or pipe system (including flanges, valves etc.,), that requires minimal work to install in the field. That is, work is limited to completing joints where the assembly connects with the existing facility – all other welding / flanging is completed prior to moving to site.

SSPC: See the *Society for Protective Coatings*.

SRB: See *sulfate reducing bacteria*.

stabilizing toe berm: See *slope grading*.

standard deviation: A statistical measure of the spread of data set relative to its mean. That is, a low standard deviation value indicates that the majority of data points are within a relatively narrow range around the mean whereas a high standard deviation value mans that the data points are distributed in a wider range relative to the mean. Technically, the standard deviation is calculated as the square root of variance for a given type of probability distribution (see various entries for probability distributions). For a specific sample of data, the standard deviation can be calculated as follows:

$$S_x = \sqrt{\frac{1}{N} \sum_{i=1}^{N} (x_i - \bar{x})^2}$$

where S_x = standard deviation of the sample data set
N = number of data points in the sample
x_i = value of the specific data point
x = mean of the sample data set

standard electrode potential: The potential associated with any reference electrode used in taking measurements associated with cathodic protection systems. The most commonly used reference electrodes for cathodic protection systems are copper-copper sulfate $(Cu-CuSO_4)$.

standard iron bar: A monument that is 25 mm square and 1.2 m in length, usually used to indicate a legally surveyed line or boundary.

stand-off: The separation of the transducer from the test specimen (or pipe wall) by a water path (or some other couplant) during ultrasonic testing. More specifically, optimisation of the amount of stand-off in ultrasonic pigs for crack detection is a key factor. Further, with stand-off, the sensors do not directly contact the pipe wall and are less prone to damage during inspection.

start defect: Inadequate fusion of the weld metal with the parent material, mitre pass, LOF, or porosity at the beginning of a weld bead.

station discharge: A point along the pipeline that is on the high-pressure side of the compressor/pumping station boundary where the fluid moves into the main-line pipeline(s) for transport.

station suction: A point along the pipeline on the low-pressure side of the compressor/pumping station boundary, where the fluid moves from the main-line pipeline(s) into the station for compression/pumping.

station yard: The area associated with a compression or pumping station that contains all of the supporting equipment and piping; typically, clearly indicated by fencing used to define the area.

stick welding: An informal term for shielded metal arc welding (SMAW). See *weld.*

stone damage: Damage done to the pipe coating, pipe material, or both as a result of rocks in the immediate vicinity of the pipe. The damage may occur during the construction process, where stones in the ditch or backfill cause damage through impact. Alternatively, issues may result over the long term if the pipe is resting on/against rock material, and stresses cause damage as the stone impinges on the line causing deformation.

stop defect: Inadequate fusion of the weld metal with the parent material at the end of a weld bead.

Stopple fitting: Used to establish a temporary bypass for a pipeline section while it remains in operation. Specifically, it is the step involving the insertion of a temporary plug into the line through a hot tap tee. See *hot tap*.

strain: The amount of deformation a material experiences due to some applied stress. A full description of the strain on a small element is a 3-by-3 tensor. Strictly speaking, it is non-dimensional deformation, usually with respect to an original length. Also, thermal strain can occur independent of any stress. The reader is referred to *Stress Strain Curve* entry for a more detailed discussion of the topic.

strain-based design: A specific variant of *limit states* design. Used when longitudinal loads that will cause permanent deformation, i.e., plastic strain, is expected.

strain gauge: Used to measure deformation on a pipeline section or material sample. They are usually simple resistive devices that measure the length and cross section of the change with applied strain. The device typically is attached directly to the pipe surface, requiring a strong, inelastic, bond. The most common application for pipeline integrity purposes is for the monitoring and measurement of pipe deformation resulting from ground movement or other geotechnical mechanisms.

strain hardening: A phenomenon where material strength increases (to a point) as it undergoes plastic deformation. The extent and nature of this behavior is a result of the specific microstructure and can be varied widely in the manufacturing process to produce more desirable material properties. See *stress-strain curve*.

strain rate: Refers to the amount of strain per unit time where strain is defined as the amount of deformation, say in length, divided by the original length. Specifically, in materials testing strain rate refers to the amount of strain that a sample is subjected too per unit time. Alternatively, it can refer to the rate at which a sample deforms when subjected to a specified stress level.

strain relief: 1. Encompasses the range of processes used to reduce or eliminate any residual stresses, detrimental to fatigue life, in a metal component. The most common of these processes is the use of heat treatment to return the microstructure to a more desirable state after machining, grinding, welding, cold or hot working of the metal. The most common form of mechanical stress relief process is *shot peening*. **2.** Strain relief is the re-sloping of a hill to reduce the stress from soil movement on the pipeline.

stratified flow: A form of *multiphase* flow where two or more phases largely separate into layers as they flow along the length of the pipeline. In this *flow regime*, lower density fluids flow in the upper portions of the pipeline. The flow within any one of the layers may be either *laminar* or *turbulent*. Figure 82 shows this flow regime.

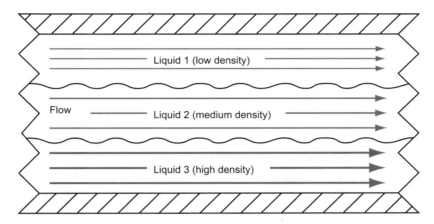

Figure 82. Stratified flow. (10)

stray current: A specific form of electrical interference of the pipeline and its cathodic protection system where there is apparently random, sustained or transient, flow of electrical current on a pipeline caused by the interaction of the cathodic protection system with other sources of electrical energy, such as cathodic protection systems from other pipelines, power lines, electric trains/trams or telluric (natural) activity. This interference results in the transient flow of current in a path or manner other than that intended. The interaction creates random spikes in pipe-to-soil potentials and can significantly impact the integrity of the line depending on the nature of the interaction. With the advent of more modern equipment, there are techniques for detecting and filtering the impact of stray currents on pipe-to-soil readings.

Mitigating stray currents is important, especially where cathodic protection is used in the vicinity of unprotected structures. Several mitigation techniques are available depending on the nature and magnitude of the stray current (to the extent they are predictable and/or the interfering structure can be identified), electrical bonding of the two structures, installation of additional anodes, and the use of coatings or shields.

stray current corrosion: A relatively uncommon situation where highly localized and deep corrosion occurs where a coating defect, combined with the discharge of stray current from the pipe to the soil or an unprotected structure, results in abnormally high corrosion rates (up to 10 mm/year have been observed). See *stray current.*

stress: The force intensity that a material is subject to (i.e., the force per unit area). See Figure 83. Stresses include pressure and tangential forces.

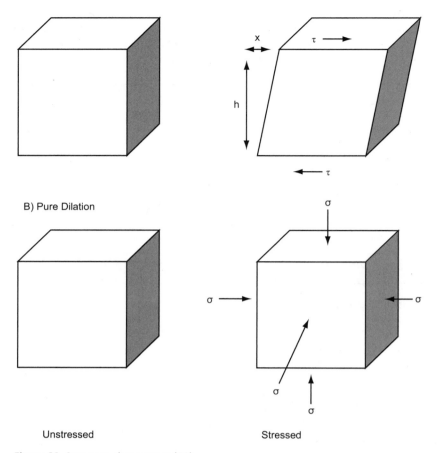

Figure 83. Stress vs. shear stress (23).

stress analysis: The engineering specialization focused on assessing the specific loads, forces, and stresses to which a structure (or structures) is subjected, with the ultimate goal of determining if the equipment can be operated in a safe and effective manner. The forces may be static, dynamic, or cyclic. In the context of pipeline integrity, stress analysis concepts are applied to wide range of situations, such as the application of soil loading on a slope to station yard piping subject to equipment vibration.

stress concentration: Caused by a discontinuity in a structure, a defect or a geometry change that results in significant increase in localized stress. The localized effect, therefore, increases the susceptibility of the material to overload, fracture, fatigue, corrosion, and cracking mechanisms. For this reason, often shallow cracking is "buffed" out of the pipe surface as a mitigating measure. Figure 84 shows the impact of geometry changes on flow stress within the pipe wall.

stress corrosion cracking (SCC): A form of Environmentally Assisted Cracking (EAC) that occurs at stresses below the ultimate tensile strength due to the combined action of stress, a corrosive environment, and the properties of the material. Tables 83 and 84 show factors and categorizations of SCC.

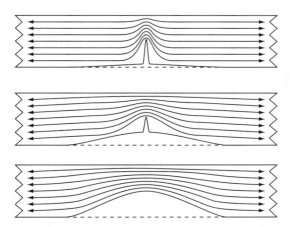

Figure 84. Geometry (i.e., stress concentrator) impact on flow stress in pipe wall (25).

Table 83. Key conditions/factors for the occurrence of SCC (25)

Factor	Description
Stress	Stress may be applied, residual, or both Stresses are constant or slowly varying (May take several hours to many years to produce failure)
Corrosive Environment	Environment only mildly corrosive except for corrosion-resistant alloys
Susceptible Material	Susceptibility of the alloy depends on composition, processing, history, & microstructure

Table 84. Categorization of SCC (24)

Category	Description
I	SCC features with a failure pressure ≥ 110% x MOP x company defined safety factor (failure pressure typically equating to 110% of SMYS)
II	SCC features where 100% x MOP x company defined safety factor (failure pressure typically equating to 110% of SMYS) ≤ failure pressure < 110% x MOP x company defined safety factor (failure pressure typically equating to 110% of SMYS)
III	SCC features where MOP < failure pressure < MOP x company defined safety factor
IV	SCC features with failure pressure equal to or less than MOP

Stress-oriented hydrogen-induced cracking (SOHIC): A form of hydrogen-induced cracking where tensile stress causes individual HIC ligaments to form in a stacked, through-thickness array. The array is oriented perpendicular to the principal applied stress. Linking up of the cracks then results in a through-wall defect and loss of integrity. SOHIC is most commonly found in weld heat-affected zones.

stress relief: The range of thermal and mechanical processes, usually applied at the end of the fabrication process, used to reduce or eliminate any residual stresses, detrimental to fatigue life, in a metal component. The most common of these processes is the use of heat treatment to return the microstructure to a more desirable state after machining, grinding, welding, cold or hot working of the metal. The most common form of mechanical stress relief process is *shot peening*.

stress-strain curve: The definition and evaluation of several mechanical properties for pipeline steels based on the results of testing their performance under tensile, uniaxial, uniform loading. The results of this testing are plotted on a stress-strain curve. Two systems can be used for plotting stress-strain diagrams; one based on *engineering* stress and strain and one based on *true* stress and strain. The engineering system is based on the original dimensions for the sample and as a result can be inaccurate (particularly for ductile materials) at strains above ~10%. The alternative system is based on the instantaneous dimensions of the sample. Figure 85 shows typical stress-strain diagram, using both systems for a typical low carbon steel.

For stresses below yield stress, the deformation is *elastic* (i.e., the material returns to its original dimensions once the stress has been removed), and the relationship between the applied stress and resulting strain is essentially linear. Once the yield point has been reached, any additional deformation is *plastic and elastic* and is permanent – that is, the sample will not return to its original shape. It should be noted that the yield point is not always clearly defined for ductile materials, but most mild steels do exhibit a sharp yield point due to the nature of the microstructure. If the stress is increased past the yield point, there is a period of strain hardening, and the specimen begins to uniformly increase in length and decrease in X-section area. Eventually, a condition is reached (UTS) where uniform elongation cannot be sustained, and the specimen begins to "neck." At this point strain accumulates rapidly and X-section stress increases even though the cross-sectional area of the specimen is smaller.

It should be noted that some higher strength steels may not exhibit such a well-defined yield point. In these situations, a "proof strength" or "offset yield strength" is established as a measure of material strength. Specifically, a test piece is stressed to produce a small specified amount of plastic deformation (usually 0.2%). The strength value is then established by drawing a line parallel to the elastic portion of the stress/strain curve at the specified strain. Thus, "offset yield strength" is a function of the specified strain value – it is not a material constant.

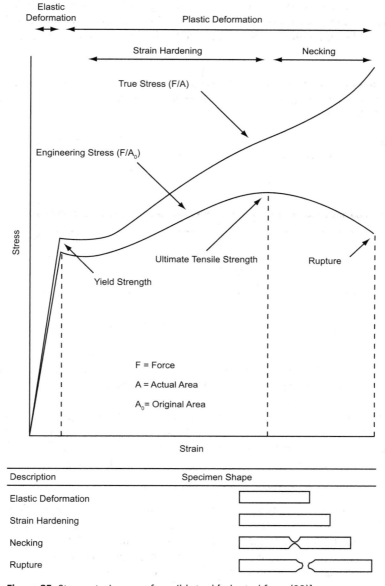

Figure 85. Stress-strain curve for mild steel [adapted from (23)].

strict anaerobes: A large group of bacteria that thrive in environments that do not have any oxygen present – they may even be harmed by the presence of oxygen.

submerged arc welding (SAW): See *weld*.

subsidence: The overall reduction in the height of ground level over time due to geological effects. Various causes can produce the overall lowering of the surface level including, but not limited to, thaw settlement, subsurface erosion (typically caused by the movement of groundwater), and karst collapse (where underlying limestone is dissolved by drainage or groundwater over time).

subsurface erosion: The erosion of material below the right-of-way surface by groundwater. The phenomenon can cause loss of support below the pipe and void development, which could result in the line being subjected to unexpected loading and stresses.

sulfide stress cracking: A specific form of hydrogen stress cracking where the presence of H_2S (i.e., in sour environments) encourages the adsorption of hydrogen into the pipe material. See Figure 86.

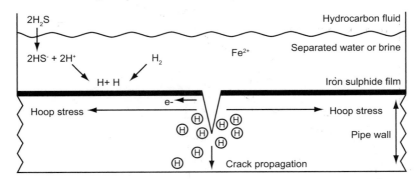

Figure 86. Mechanism of sulfide stress cracking [adapted from (25)].

Sulfate-reducing bacteria (SRB): A large group of anaerobic organisms that use sulfate, sulfur, or other sulfur compounds, such as sulfite and thiosulfate as oxidizing agents. In the context of pipeline integrity, these bacteria can play an important role during the microbiologically-influenced corrosion processes. SRB can grow under the pipeline coating, causing a depolarization effect and increasing the corrosion rate, even if the pipeline is cathodically protected (the bacterial action forms a dielectric film that shields the pipe from the effects of the cathodic protection system).

super duplex steel: A sub-class of duplex steels that is often specified for use in situations where higher corrosion resistance is required (such as sour service). These steels usually have specific microstructure and alloying requirements as determined by a *pitting resistance equivalent number (PREN)*. Specifically, super duplex steels have a PREN value greater than 40. Further guidance can be found in NACE MR0175. See *pitting resistance equivalent number* and *duplex*.

supervisory control and data acquisition (SCADA): Describes the communication, instrumentation, and computer software systems that control pumps, valves, and other pipeline components. It allows rapid detection of major incidents without requirement for regular inspection and gives the control center the capability to remotely start or stop certain compressors/pumps, thereby changing flow volumes to meet changes in customer demand for natural gas/oil. SCADA systems on pipelines have two distinguishing features from other industries: they cover a long geographical distance and they may cross differing regulatory regimes that may have differing operating requirements. It should be noted that although most transmission systems are equipped with a SCADA system, having such a system is not in itself a regulatory requirement. Table 85 shows the main components of a SCADA system.

Table 85. Key components of a SCADA system (10)

Component	Description
Master Terminal Units	Also known as MTUs, this is the central host usually collocated with the control center
Remote Terminal Units	Also known as RTUs, these are the remote (field) data gathering and control units
Communication Software System	The link between the MTU and RTU's to organize the analysis and display of data in the computer screen (or printer or other output medium) and remotely control field equipment connected to the SCADA system (e.g., valves, pumps, etc.)

surface defect: Any shallow defect in the pipe material that breaks through the outside of the pipe wall. The term encompasses several types of defects including, but not limited to, surface laps, slag inclusions, slivers, transverse tears, and surface roughness.

surface finish: The physical state of exposed pipe metal based primarily on texture, surface mechanical properties, and visual appearance. NACE and the Society for Protective Coatings (SSPC) have developed standards in this area. In the context of pipeline integrity, surface finish is critical for the application of protective *coatings* and for the effective use of non-destructive techniques, such as *magnetic particle inspection* or *liquid particle inspection*. A list of relevant standards appears under the term *surface preparation*.

surface lamination: A break in the pipe wall that generally lies parallel to the surface of the pipeline and is characterized by the presence of non-metallic impurities. Surface laminations are defects that result from the rolling out of inclusions or blow holes during the plate manufacturing process. While not all laminations pose an integrity threat, surface-breaking laminations can serve as initiation points for corrosion.

surface preparation: Includes the methods and procedures that must be undertaken to meet a specified physical state of exposed pipe metal based primarily on texture, surface mechanical properties, and visual appearance. NACE and the Society for Protective Coatings (SSPC) have developed standards in this area – in some cases the standards have been developed jointly. Table 86 lists the most relevant standards.

surge pressure: The level of (over)pressure produced by a rapid change in velocity of the moving stream that results from shutting down a pump station or pumping unit, closure of a valve, or any other blockage of the moving stream.

swabbing: Refers to the removal of foreign material from a pipeline typically through the use of cleaning or bi-directional pigs.

Table 86. Surface preparation standards most relevant for pipeline integrity related work (186)

Standard	Title
• NACE No. 1 • SSPC-SP 5	White Metal Blast Cleaning
• NACE No. 2 • SSPC-SP10	Near-White Metal Blast Cleaning
• NACE No. 3 • SSPC-SP6	Commercial Blast Cleaning
• NACE No. 4 • SSPC-SP7	Brush-Off Blast Cleaning
• NACE No. 5 • SSPC-SP12	Surface Preparation and Cleaning of Metals by Water Jetting Prior to Recoating
• SSPC-SP 1	Solvent Cleaning
• SSPC-SP 2	Hand Tool Cleaning
• SSPC-SP 3	Power Tool Cleaning

swamp mats: Square, flat structures used to build temporary pathways for heavy equipment in areas where the ground is soft and/or wet. While exact design and construction varies, the principle remains the same in that the mats are used to distribute the weight of the equipment over a larger area so that the soft soil can support the load (i.e., the equipment is less likely to become stuck). Swamp mats also protect fragile top soil and vegetation.

swamp weights: See *saddle weights.*

sweet crude: Crude oil that contains less than 1% total sulfur content (usually in the form of H_2S). Handling of sweet crude is typically easier than sour crude because it is less toxic to humans and less corrosive of metals.

sweet gas: Natural gas where the sulfur content is too low to be measureable. Handling sweet gas is typically easier than sour crude because it is less toxic to humans and less corrosive of metals.

sweet service: When a pipeline system carries product with sulfur content that is less than 1%. Thus, no specific design, construction, or operational requirements associated with sulfur or H_2S handling and containment are required.

T

tack coat: A layer of adhesive material applied to the pipeline surface to ensure that (cold-applied) tape coating adheres effectively. See *coating*.

Tafel equation: An equation that relates to the rate of an electrochemical reaction (as the current density) to the corresponding over-potential (i.e., losses or energy required for the reaction above and beyond the levels predicted by thermodynamics). The equation is as follows:

$$\Delta V = A \times \ln\left(\frac{i}{i_o}\right)$$

where ΔV = over-potential, V
  A = "Tafel Slope" (determined experimentally), V
  i = current density, amps/m^2
  i_o = exchange current density, amps/m^2

tap: A small diameter *tee* or off-take used to deliver product to a customer (or customers) directly from the main-line.

tape coating: Refers to the range of external pipeline coating formulations that are applied in a narrow strip of material wrapped around the pipe – regardless of the constituent compounds. See *coating*.

tee: A fitting that allows two pipeline sections to interconnect. The connection occurs in the shape of the letter "T," such that the "branch," or connecting pipe, is at 90° to the straight section allowing two inflows to combine into a single outflow (or alternatively, a single inflow to result in two outflows). The branch connection maybe of the same diameter as the straight section of pipe or have a diameter that is significantly smaller.

In the context of pipeline integrity, the size, orientation, and the presence of cross bars on the branch section becomes critical for in-line inspection operations. Specifically, branches located in the 6 o'clock position pose a greater threat to inspection tools (vs. those oriented at the 12 o'clock position) because gravity has a tendency to "pull" them into the branch – this effect is further exacerbated if there is also product flowing into the branch (as opposed to flowing into the straight pipeline section). This concern is also directly related to the size of the branch relative to the straight pipe section (size-on-size connections result in the greatest risk of tool damage). To ensure a line is piggable, most branch connections that have a diameter that is ≥30% of the straight pipe diameter are installed with cross bars (also known as pig or scraper bars).

telluric currents: A specific type of stray current caused by natural phenomenon, such as geo-magnetic forces, tidal currents, and sunspots. These currents can create random, transient changes in the pipe to soil potential that can disrupt cathodic protection and affect pipeline integrity. The detection and management of telluric current is similar to the methods and techniques used for stray currents in general. See *stray currents*.

temperature recorder: A device used to capture temperature readings – most often used in the context of hydrostatic testing of pipelines where the water temperature can significantly impact test conditions.

temporary repair: A short-term restoration of pipe structure to ensure that the immediate integrity of the pipeline is maintained; often undertaken in situations where it may not be practical to remove the line from service or where other conditions are a barrier to the completion of a more permanent repair.

tensile strength: A property of pipe that is the stress at the maximum of the stress-strain curve for the material (see *stress-strain curve*). This term is often used interchangeably with the term *ultimate tensile strength*.

tenting: Refers to the phenomenon where tape-coated lines have a significant gap between the coating and the pipe wall in the area adjacent to the long seam or spiral weld. See Figure 87. This can be an integrity threat if water enters the gap as the area becomes particularly susceptible to corrosion (see *narrow axial external corrosion*), especially because tape coatings can impede cathodic protection.

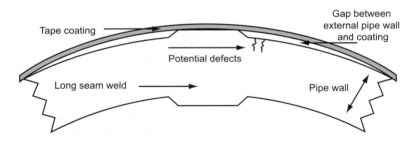

Tenting over long seam weld

Figure 87. Tenting phenomenon (10)

terrain: Macroscopic environmental conditions to which a pipeline is subject as a result of the route selection. While conditions will vary to some degree on a meter-by-meter basis, terrain refers to the general characteristics of the surface features, ground conditions, topographic relief, hydrology, and climatic effects. Terrain is particularly important for initial pipeline route selection, but is also important for pipeline integrity, especially as it relates to geotechnical hazard management. Table 87 cites nine main types of terrain, and associated descriptions, relevant for pipeline route selection.

Table 87. Nine major terrain types [adapted from (26)]

Type	Description
Glaciated	• Areas that exhibit characteristics formed by the melting and retreating of continental ice sheets and glaciers. The landscapes tend to comprise a range of stratified water-borne deposits (till). The grain size of these deposits can vary significantly (i.e., sandy to boulder size).
Fluvial	• Landscapes that are largely a product of long-term exposure to running water (rivers) resulting in areas of significant erosion and/or accretive deposition over time.
Permafrost affected	• The strict definition of permafrost is where the soil or rock temperatures remain below 0°C for two consecutive winters (including the intervening summer). Any thawing/freezing cycles can have a significant impact on drainage, buoyancy, and stability of the pipeline right-of-way. See also *permafrost*.
Peatland / wetland	• A grouping that includes swamps, marshes, bogs, and fens. While further classification of subtypes has been undertaken, key characteristics include landforms that are waterlogged, highly compressible, and tend to occur in cool climates where there is poor drainage of rain and snowmelt. While these areas are water saturated most of the time, the water table can drop dramatically during the hot season or during prolonged drought.
Mountain	• Terrain, typically along continental margins, that includes mountain ranges as well as the intervening valleys, lakes, and basins. The high-relief slopes, often combined with high-precipitation rates, result in particular concerns associated with slope failures.
Volcanic	• While pipelines are unlikely to be routed in active volcanic regions, currently dormant regions can also pose significant challenges including steep slopes, highly plastic clay surfaces, networks of open fissures, seismic activity, landslides, etc.
Coastal	• Coastal environments undergo constant change due to wave action and ocean currents. The effects of wave action are typically isolated to +/-30ft/10 m below sea level, but the effects of ocean waves and current can often be greater in terms of shore recession/advancement.
Karst	• A topography characterized by sinkholes that result from significant surface and subsurface water flow, dissolution, subsidence, and underground cavern collapse in carbonate and evaporate strata.

test coupon: A metal sample buried in close proximity to the pipeline to assess the specific conditions (such as corrosion rate) which the pipe experiences. While the use of test coupons can be valuable in terms of providing input data for detailed analysis (e.g., remaining life calculations, corrosion growth analysis, etc.), a relatively long time frame, and multiple coupons, are required to obtain meaningful information.

test failure: The failure of a pipeline section during a hydrostatic or *pressure test*.

test head: An assembly attached to the end of a length of pipeline to facilitate *pressure testing*. A test head installation can be a modified launcher or receiver with valved connections necessary for filling, pressuring and instrument lines. The pipeline section to be tested will have the test heads at each end, and the section is usually filled with water using pumps which have the capacity to overcome pressure due to the static head of the pipeline.

test lead: See *test point.*

test point: A wire, enclosed in an above-ground box or housing, that is connected to a buried pipe. These wires are typically placed at regular intervals on the pipeline and used for routine monitoring of cathodic protection levels.

test pressure: The final high-point pressure that is reached during a hydrostatic (or other pressure) test.

tethered tool: A class of tools physically connected to a cable or tether during operation. The cabling is used for various functions, including tool retrieval and power and data transmission.

thaw settlement: 1) of the pipe: A geotechnical hazard where the pipe may be subjected to significant stresses and strains caused by differential settlement of the ground that occurs due to thawing permafrost. Significant stresses and strains on the pipe can also occur at the interface of frozen and unfrozen areas (particularly in areas of discontinuous permafrost). **2)** of the ROW: A geotechnical hazard where subsidence occurs along the ROW due to the permafrost thawing. This subsidence can result in exposure of the pipe as well as disruption of surface water drainage.

thermoplastic liner: A *liner* constructed from thermoplastic material. The main difference from other liners is that the liner material softens at higher temperatures and hardens at cooler temperatures. Thus, installation methods may vary (i.e., may be heated prior to extrusion into the interior of an existing pipeline) and these liners may not be suited to high temperature applications. The resulting relatively soft tube is then expanded and, upon contacting the internal wall of the pipe, cools and hardens. See *liner.*

third-party damage: Damage to a pipeline facility by an outside party (i.e., other than the operator or those performing work for the operator). Pipeline operators have developed various tactics to mitigate the risk associated with this type of damage including, but not limited to, aerial survey, ROW signage, "one-call," or "call before you dig" systems, indicator tapes, or even concrete slabs.

threat: An event, mechanism or condition that, should it occur, will have a negative effect. Examples of common threats on pipelines include, but are not limited to, external corrosion, internal corrosion, stress corrosion cracking, manufacturing-related defects, welding/fabrication-related defects, equipment failure, third-party/mechanical damage, incorrect operations, weather-related, or outside forces. The most common threats and associated mitigation methods appear in Table 88, where "R" indicates a largely reactive measure and "P" indicates a largely proactive measure.

three-layer polyethylene (3LPE): See *coating* (specifically Table 16).

three-layer polypropylene (3LPP): See *coating* (specifically Table 16).

through-transmission technique: An ultrasonic testing method where two transducers are required – one of either side of the material being tested. The technique relies on the change in the amplitude of the ultrasonic signal between the transmitter and receiver for detecting and sizing defects. See *ultrasonic testing* for a more in-depth discussion of the topic.

time-dependent failure mechanisms: Threats that pose an increasingly larger risk to the pipeline system as time passes. Examples of these types of mechanisms include, but are not limited to, corrosion and fatigue.

time-independent failure mechanisms: Threats that do not progressively worsen over the passage of time; they simply either occur (or do not occur). Examples include third-party damage, weather-related events, and human error.

time-of-flight diffraction (TOFD): A mechanized ultrasonic measurement method most recently applied to the inspection of girth welds (in place of traditional radiographic techniques). Key difference relative to traditional ultrasonic testing is that it accounts for energy diffracted from a discontinuity as opposed to relying purely on the amount of energy reflected back from a discontinuity.

toe crack: Cracking that occurs at the junction of the pipe surface and a weld bead.

toe erosion: A geotechnical hazard whereby the base of a slope is eroded, typically as a as a result of water flow, undermining the stability of the slope by reducing the constraint against downward movement of the soil.

TOFD: See *time-of-flight diffraction.*

tool accuracy: See *accuracy.*

top coat: The outside layer in a multi-layer pipeline coating system. See *coating.*

top hat cracking: The solidification cracking that can occur in the weld bead at the point where there is a change in profile at the weld fusion boundary. A poor weld bead profile can lead to the concentration of strain during weld bead solidification. This type of defect is best addressed through strong quality control (i.e., choice of welding parameters, weld bead geometry, high purity pipe and quality welding materials).

topographic survey: A survey of a land area in three dimensions providing information regarding the elevation change that occurs in an area. Topographic surveys can be particularly important for pipeline route selection as well as identification and assessment of potential geotechnical hazards the pipe may be exposed too.

toughness: The ability to resist fracture (i.e., a propagating crack), which can be characterized in a Charpy or Drop Weight Tear Test. Toughness is one of the key considerations in pipeline design because it is one of the primary methods of establishing *fracture control* (i.e., failure at stresses below yield is largely dependent on the toughness of the pipe material). The most common measures of toughness are the *Charpy test* and the *drop-weight tear test (DWTT)*. It should be noted that temperature also impacts the behavior of materials (i.e., they tend to be more brittle at lower temperatures) and as such, a temperature value must

tracing: See *rubbing*.

transducer: A device for converting energy from one form to another. For example, in ultrasonic testing, a transducer converts electrical pulses to acoustic waves and vice versa.

transformer rectifier: See *rectifier*.

transgranular SCC: A form of SCC associated with near-neutral pH electrolyte (typically pH 6 to 8) in which the crack growth or crack path is through or across the grains of a metal. This form of cracking has limited branching and is associated with some corrosion of the crack walls and sometimes pipe surface, leading to cracks that are typically wider than those associated with classical (intergranular SCC). Cracks are often associated with the presence of pasty $FeCO_3$ corrosion product and a dilute carbonic acid (i.e., contains CO_2). Transgranular SCC is also referred to as *near-neutral*, *non-classical*, and *low pH SCC*.

Table 88. Major controllable threats and mitigations for an existing pipeline [adapted from (8)]

Threat Type	Pipeline Patrol/ Survey of ROW	Intelligent Pigging	Product Quality	Slope Monitoring	CP/ Coating Surveys	Hydrostatic Testing
3rd party damage	P	R				R
Coating defects					P	
Construction defects		R				R
Cracking		R				R
Excessive strains /stresses (mechanical damage)	R	R		P		R
External corrosion		R			P	R
Internal corrosion		R	P			R

transmission line: A pipeline, other than a gathering or distribution line, that transports gas or liquid from a gathering or storage facility to a distribution center or storage facility; operates at a hoop stress of 20% or more of the specified minimum yield strength of the pipe; or transports gas within a storage field. Other characteristics are that most transmission pipelines are constructed from steel in dedicated rights-of-way. Further, in the case of gas pipelines, the product is not odorized and in terms of failure modes, the potential for leak and rupture exists (i.e., the failure mode for gas distribution systems is primarily leak only).

As such, the pipeline integrity programs associated with transmission systems can be quite different from those that would be considered appropriate for gathering or distribution systems. The key differences such as product quality, diameter, and consequence of failure can drive vastly different decisions in the area of inhibitors, inspection programs, and repair-vs.-replace decisions. See also *pipeline*.

transverse field MFL: A technique used on in-line inspection tools where the magnetic field is introduced in the circumferential direction into the pipe wall (vs. the more common technique of using a field in the axial direction). The alternative orientation results in greater sensitivity to long, narrow axially-oriented corrosion as the defect results in a greater "disturbance" to the magnetic field.

transverse tears: Crack-like features on the pipe surface running in the hoop (circumferential) direction. The defects can result from a situation where insufficient material is removed from the surface of an ingot so that defects like shrinkage cracks are carried on through the pipe-making process. Small tears typically do not pose a structural concern to the pipe; however, they can cause coating holidays as well as act as initiation points for corrosion activity.

transverse wave: See *shear wave* and *ultrasonic testing* for a more in-depth discussion of the topic.

trap: See *pig trap.*

trapped slag: Impurities that become integrated between individual passes of a weld run resulting in discontinuities and holes in the weld surface. The defect typically results from a current that is too low or when the welding technique is too rapid.

tri-axial sensor technology: A specific configuration of magnetic flux technology used on in-line inspection tools. Specifically, tri-axial sensor technology allows the measurement of the magnetic field in all three dimensions (x, y, and z) allowing more accurate identification and sizing of pipeline anomalies.

trunk line: The main-line, as opposed to a feeder line, of a pipeline system. It is the portion of the pipeline system that connects directly to a sales point or the pipeline system with another operator. This is typically the section of pipe with the largest diameter within the network.

tundra: A landscape where tree growth does not occur and vegetation growth, if present, is limited to low growing species due to a short growing season and permanently frozen subsoil. There are three main types of tundra: Arctic, Antarctic and alpine (i.e., due to altitude). In the context of pipeline integrity, the existence of permafrost and permanently frozen subsoil dictates that the pipeline design incorporates the relevant geotechnical considerations such as subsidence, frost heave and thaw settlement. See *geotechnical threat*.

turbidity: The degree to which a fluid is "cloudy" (i.e., lacks of transparency). The most common example and usage of this term is in reference to water that, although normally clear, can pick up particles that render it translucent or opaque depending on the level of the suspended solids. The most commonly used standardized units to measure turbidity are FTU (Formazin Turbidity Unit) or FNU (Formazin Nephelometric Units).

turbulent flow: A type of flow of a fluid, which is characterized by rough and chaotic flow. In turbulent flow, secondary random motions are superimposed on the principal flow, and there is an exchange of fluid from one adjacent sector to another. More importantly, there is an exchange of momentum so that slow-moving fluid particles speed up and fast-moving particles give up their momentum to the slower moving particles and slow down themselves. Note that the velocity distribution in turbulent flow is more uniform, but the key implication for pipeline design is that turbulent flow inherently is less efficient than laminar flow and results in a greater pressure drop over the same distance as more energy is lost through the secondary random motions. Nonetheless, it would be unecnomical to operate any transmission pipeline system in anything other than a fully turbulent flow regime, since the flow rates in a laminar regime are much too low to be practical. Figure 88 illustrates laminar vs. turbulent flow. Typically, the term is only used in the context of single-phase flow – not *multiphase flow*. For a discussion of multiphase flows, see *flow regime*.

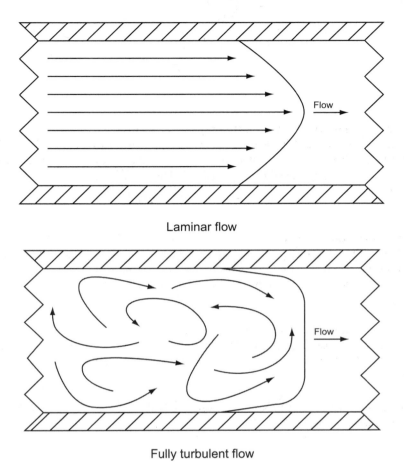

Laminar flow

Fully turbulent flow

Figure 88. Single-phase laminar and turbulent flow in a pipeline. (10)

turbulent flow erosion: The erosion and often very localized metal loss resulting from the disturbance, usually in the form of constriction, of product flow. The constriction of product flow results in higher velocities and turbulence; if the constriction is sufficient, cavitation may occur as a result of the pressure drop. Metal loss can occur at a very rapid rate and is exacerbated by the presence of particles and/or mixed phase flow (as opposed to gas phase alone). Scale and corrosion inhibitors typically present an insufficient barrier for this mechanism.

U

ultimate (tensile) strength (UTS): One of the properties used to specify the properties of pipe. It is the maximum stress that a material can withstand in a standard uniaxial tensile test without fracturing and is indicated by the stress at the maximum of the stress-strain curve for the material (see *stress-strain curve*). This term is often used interchangeably with *tensile strength*.

ultrasonic crack detection: The use of *ultrasonic testing* methods to identify the presence of pipeline cracks and/or size them. The technique may be applied manually or mounted on an in-line inspection tool – as a result, the effectiveness of the technique will vary depending on the specifics of the circumstances under which it is used (see *ultrasonic testing*).

The current state of the art provides for the detection but only limited sizing of capabilities of cracks. Most vendors report the cracks in broad depth categories of $0 - 12.5\%$, $12.5\% - 25\%$, $25\% - 40\%$, and greater than 40% of wall thickness.

ultrasonic inspection: The use of ultrasonic testing methods to assess the state of a pipeline. See *ultrasonic testing*.

ultrasonic testing (UT): The use of high-frequency sound waves above the audible range, ~20 kHz, to determine the integrity of material. Sound travels by the vibration of the atoms and molecules present, traveling with a velocity depending on the acoustic properties of the medium. Thus, imperfections and inclusions in solids cause sound waves to be scattered, resulting in echoes, reverberations, and a general dampening of the sound waves, which can be measured and interpreted. Because gases are relatively poor transmitters of ultrasonic waves, most pipeline applications require the use of the liquid couplant to ensure efficient transmission of the wave from the transducer to the bulk material. Thus, this technique has significant challenges in the internal inspection of gas pipelines where the inspection tools must be run in a liquid batch or slug to facilitate inspection.

There are two main types of waves to be considered in the context of ultrasonic technology: transverse waves (also known as shear waves) and longitudinal waves (also known as compression). Transverse waves are characterized by the motion of atoms and molecules in the bulk material moving in a direction *perpendicular* to the direction of the wave propagation, whereas longitudinal waves are characterized by atoms and molecules in the bulk material moving in a direction *in-line* with the direction of the wave propagation.

Figure 89 illustrates this concept. The type of wave used to undertake testing is important as there are several trade-offs in choosing longitudinal vs. transverse for testing – one of the key differences being that transverse waves tend to be more sensitive to smaller discontinuities (due to a shorter wave length), but do not travel in liquids and do not travel well in plastics and other couplants.

There are also several transducer/receiver arrangements that can be used in ultrasonic testing with two broad categories of pulse-echo techniques and through- transmission techniques. Figure 90 illustrates the basic arrangements. Other wave modes such as surface waves are also used for specific purposes.

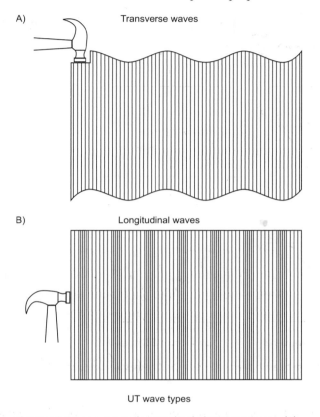

A)   Transverse waves

B)   Longitudinal waves

UT wave types

Figure 89. Transverse vs. longitudinal ultrasonic waves. (1)

ultrasonic wall thickness gauge: An instrument, relying on ultrasonic wave technology (see *ultrasonic testing*), used to determine the wall thickness of pipe.

unbarred tee: A tee fitting that does not have any cross bars on it. If the unbarred tee is of sufficient diameter (i.e., a diameter greater than ≈30% of the diameter of the straight pipe), then it may pose a risk for pigging operations. See *tee fitting*.

undercut: A linear defect at the edge of the internal or external weld bead that is the result of metal being removed by the force of the welding arc (i.e., welding current too high), or if the angle of the welding electrode is incorrect. These defects are typically rounded, and limited lengths of the defects are allowed in most specifications (assuming the depth is not significant). Where required, defects can be removed by grinding and addressed through a local weld repair.

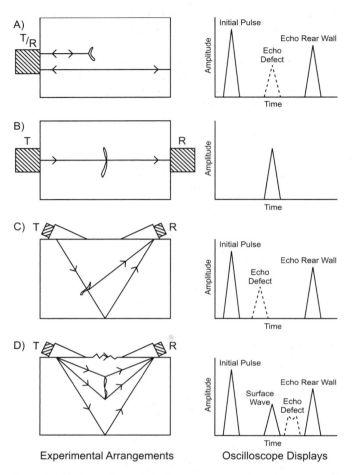

Figure 90. Various ultrasonic testing arrangements and corresponding signals (1).

under-protection: A condition where there is insufficient cathodic protection of a pipeline relative to established criteria (such as the standard -850 mV criteria).

uneven wall thickness: A level of variation in the pipe wall thickness greater than one would expect based on the tolerances associated with the specific fabrication process; usually, 5% of the specified wall thickness.

unpiggable: A pipeline section is considered unpiggable when it is not possible, or cost effective, to use an in-line inspection tool (pig) for various reasons. Typically, if constraints exist in the pipeline, one simply undertakes a program to modify the affected area(s) to ensure safe pig passage. However, if there are sufficient   restrictions, it may become cost prohibitive, or impractical, to undertake the required modifications. In this scenario, the operator is forced to pursue other methods for ensuring integrity of the pipeline. Table 89 shows some of the most common factors that result in "unpiggable" pipelines.

Table 89. Examples of constraints that could result in a
pipeline being considered "unpiggable" (124)

Constraint	Description
Tight radius Bends	• Tight radius bends could result in a tool becoming lodged. • While specifications vary significantly by tool type and vendor; generally, high-resolution inspection tools require bends greater than 3D for safe passage.
Unbarred Tees	• Tee fittings could result in a tool becoming wedged in a pipeline if the intersecting line is of sufficient diameter – especially, if product exits in the main pipeline at that point. • While specifications vary significantly by tool type, unbarred tees in the 6 o'clock position are of particular concern, as are off-takes that are greater in diameter than 50% of the main pipeline diameter.
Diameter Restrictions / Transitions	• Any restriction in the diameter of the pipeline could result in a pig becoming lodged, or significantly damaged. • Diameter restrictions can be a result of both planned and unplanned. Planned items can include dual diameter pipelines, heavy-wall pipe sections. Unplanned items can range from simple temperature/ pressure probes left in the pipe prior to a pig run all the way to pipeline deformation, such as buckling, which is an integrity concern in itself. • Typically, the more sophisticated the tool becomes, the more sensitive it is to diameter restrictions. For example, some cleaning pigs can pass through restrictions as great as 30% of the pipe diameter, whereas some high-resolution tools could be significantly damaged by restrictions as small as 7% of the pipe diameter. • Eccentric reducers at launch/receive can cause tool damage depending on orientation and the abruptness of the transition.
Lack of Launch / Receive Facilities	• Launch and receive facilities are necessary for launching pigs to ensure that an appropriate means for inserting the tool into a high(er) pressure line is available. • In some cases "temporary" (i.e., moveable) facilities can be used – although this becomes increasingly difficult as pipe diameters increase. • In some cases, factors such as line configuration or lack of physical space can limit an operator's ability to install launchers/receiver deeming the section unpiggable
Valves	• Full-bore valves are crucial for safe passage of inspection tools as valves with restrictions (e.g., plug valves) can damage tools, or in some cases trap a tool. • Typically, full-bore ball valves are best suited for pigging operations, but other valve types and configurations may also be acceptable depending on the specifics of the tool.
Inability to Control Product Flow	• Smart inspection tools must operate within a specific speed envelope to collect good quality data. • The ability to control the speed of the flow is critical for the entire length of the inspection.

UOE (pipe manufacture): Refers to pipe manufactured in a process where the plate is first formed into a U shape, then an O shape prior to joining the two edges together (usually by an internal then an external SAW weld, i.e., *DSAW*). The next step involves forming the pipe into a circular shape by cold expansion.

upstream: A term indicative of a direction along the longitudinal axis of a pipeline relative to a given reference point that may be located anywhere along the line; specifically, the upstream direction is towards the point from which the product flows.

Unusually Sensitive Areas (USA): A term used in Part 195 of US regulation governing the transportation of hazardous liquids. Specifically, this term refers to a drinking water or ecological resource area that is unusually sensitive to environmental damage from a hazardous liquid pipeline release.

U/S: See *upstream*.

USA: See *unusually sensitive area*.

US DOT PHMSA: See *(US Department of Transportation) Pipeline and Hazardous Materials Safety Administration*.

useful life extension: The process whereby equipment is serviced, and/or rehabilitated, such that it can remain in-service, safely for a significantly longer period of time. This typically refers to a set of activities that are associated with extensive refurbishment and go above and beyond routine maintenance. Numerous examples of this can be cited including, but not limited to, recoating significant lengths of an existing pipeline, re-rating of a pipeline for higher pressure service (usually through hydrostatic testing) or extensive repair programs following in-line inspection.

UT: See *ultrasonic testing*.

utility pig: A class of in-line inspection tools that perform a function such as cleaning, separation of batches or dewatering the pipeline. See *pig* for a more detailed discussion of the various types of utility pigs.

UV degradation: Cracking and embrittlement of coatings due to ultraviolet radiation exposure during prolonged outdoor storage or service above ground.

V

valve: A device that assists with the control of fluid flow in a pipeline. Valves can serve many functions and, as a result, can be specified in various configurations. The main valve configurations used in hydrocarbon pipelines, and their primary functions, appear in Table 90. Basic schematics of the various valve types are available under the individual entries. One of the most important functions of valves in the context of pipeline integrity is the ability to facilitate the isolation of a pipeline if the section has failed or during inspection, excavation, or repairs are being undertaken.

valve operator: The mechanism used to adjust the position of a valve (regardless of valve type). There are two basic types of mechanisms – mechanical (including gear-operated) and powered -- and many variations within these two basic categories, depending on the situation. Factors affecting the choice of operator type are based on, but not limited to, the size of the valve, type of valve, type of service (i.e., station bypass vs. automatic line break control), frequency of use, access for maintenance, requirement for remote operation, and duration of service in the context of any of these factors.

velocity excursion: A period during which the speed of an in-line inspection tool exceeds the limits for quality data collection. See Figure 91. In-line inspection tools have a speed range in which several parameters are optimized – the most important of which is the quality of data collection used to identify anomalies. During a velocity excursion, there is potential for data degradation as well as physical damage to the tool. The optimal speed range for in-line inspection tools varies by vendor, technology, and model.

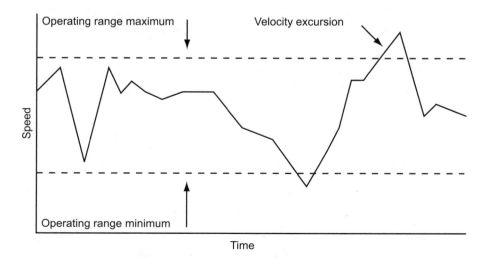

Figure 91. Example of in-line inspection tool velocity plot with speed excursion.

Table 90. Major valve types [adapted from (27) and (10)]

Type	Subtypes	Port Size	Typical Usage	Remarks
Gate	• Rising stem • Non-rising stem • Slap gate • Wedge gate • Expanding gate	• Full port • Reduced port • Through conduit	• Isolation (i.e., on/ off operation)	• See *gate valve*
Ball	• Floating ball • Trunion mounted ball	• Full port • Reduced port	• Isolation (i.e., on/off operation)	• Well suited to high pressure differentials • Special designs can be used for flow or pressure control • Shut-off; ideal for quick shut-off because 90° turns offer complete shut-off • See *ball valve*
Butterfly	• n/a	• n/a	• Isolation (i.e., on/off operation)off • Regulating/ throttling pressures	• See *butterfly valve*
Plug	• n/a	• Full port • Reduced port	• Isolation (i.e., on/off operation)off • Regulating/ throttling pressures	• Special plug valve designs can be used for flow or pressure control • See *plug valve*
Globe	• n/a	• n/a	• Regulating/ throttling pressures	• See *globe valve*
Check	• Clapper • Thru-conduit clapper • Lift	• Reduced port	• One-way flow	• See *check valve*

verification dig: An excavation on a section of pipeline to confirm the accuracy of the data from an in-line inspection. Specifically, the pipeline is exposed and specific defects are located, sized, and compared to the results from an internal inspection tool. Generally, a sufficient number of defects must be assessed in this manner to ensure that the results are statistically valid prior to reaching a conclusion regarding tool accuracy. The potential for an in-line inspection tool performing outside the stated specification should be addressed in advance of an inspection with clear agreement regarding the protocols to be followed to establish tool accuracy. Further, discussions between the vendor and client should extend to the implications/consequences of inspection data not meeting specified tolerances based on the root cause of the problem (e.g., lack of line cleanliness vs. tool malfunction).

Vickers hardness test: A standardized non-destructive test used to determine the hardness of a material. Hardness is a measure of a material's resistance to plastic deformation or penetration. It is an important parameter to control and assess on a pipeline because harder materials are often more susceptible to various forms of cracking. The Vickers hardness test estimates hardness by pushing the surface of a material with a diamond indenter of a specific size and shape and with a specific force. The hardness is estimated by examining the size of the indentation and can be related to *yield strength* through an empirically determined equation. See *hardness* to see a comparison of major test methods.

vinyl tape: See *coating* (specifically Table 16).

visual inspection: The systematic assessment of a section of pipeline, or an associated piece of equipment, based on physical appearance. Most often, this involves looking for obvious leaks, damage, or other evidence of non-performance/non-compliance to code. Typically, if a visual inspection reveals an anomaly, further examination is required using non-destructive techniques to determine the severity and potential repair requirements. While visual inspection is not always conclusive, it is a critical component of assessing pipeline integrity defects/issues. As such, a record of the visual inspection (i.e., photographs) is a critical component of documenting significant pipe integrity related activity.

voltage: A measure of the electrical potential energy per unit charge that would drive an electrical current between two points. That is, if a unit of electrical charge were placed in a location, the voltage indicates its potential energy at that point that results from its interaction with surrounding electrical field.

voltmeter: A device used to measure the electrical potential difference (i.e., voltage) at two different points on an electrical circuit. While less of a concern with more modern equipment, in choosing a correct voltmeter, one must consider the range of potentials expected in pipeline cathodic protection systems as well as whether they are direct current (vs. alternating current) systems.

volume balance: The process of determining the difference, if any, between the amount of fluid entering a pipeline and the amount of fluid leaving a pipeline. A significant difference between the two volumes can be indicative of a leak on the pipeline system. This method is most readily applied for (essentially incompressible) fluids because temperature and pressure differences must be accounted for. It should be noted that equipment accuracy and pipe characteristics result in a lower threshold below which small leaks will not be detectable. Further, the complexity of most pipeline systems means that the application of this concept is typically software-based. The reader is referred to API 1149: Pipeline Variable Uncertainties and Their Effects on Leak Detectability for a more detailed discussion on the topic.

wall thickness (WT): See *nominal wall thickness*.

wall thickness transition: Any location along the pipeline where the wall thickness of the pipe changes. This change in wall thickness could be correlated with several factors, such as changes in class location and road crossings. If the change in wall thickness is significant enough, a shorter piece of pipe, with a wall thickness that falls between the wall thicknesses of the two sections to be joined, is used to ensure that the pipe joints can be welded together effectively.

water blasting: See *water jetting*.

water crossing: The location(s) where a pipeline crosses any body of water (e.g., water, lake, stream, etc.,). Typically, a range of considerations specific to water crossings must be addressed during the design, construction, and operations (i.e., all stages of the pipeline lifecycle).

water hammer: See *pressure surge*.

water jetting: The use of high-pressure water for various activities; however, the primary application in pipeline systems is for surface preparation – for example prior to coating and/or inspection during excavation. See Table 91.

wavy stratified flow: A form of *multiphase* flow where a liquid and gas phase form separate layers as they flow along the length of the pipeline. However, the nature of this *flow regime* is such that the gas phase causes waves in the liquid phase at the interface.

Figure 92 shows this flow regime.

Table 91. Summary of water jetting categories based on pressure (186)

Type	Description
Low pressure	Water pressures are < 34 MPa (5,000 psig)
Medium pressure	Water pressures are < 34 MPa (5,000 psig)
High pressure	Water pressures are > 70 MPa and < 170 MPa (i.e., 10,000 to 25,000 psig)

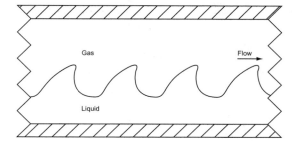

Figure 92. Schematic of wavy stratified flow. (180)

wax: A class of hydrocarbon (i.e., water insoluble) compounds that naturally occur in crude oil. Waxes have various properties – most relevant of these is the melting point that can range from room temperature to greater than 100°C.

In the context of pipelines, one of the most relevant considerations is the gradual deposition of wax inside the pipeline that increases with a lower temperatures and lower flow rates.

Wax deposits can adversely affect the accuracy and performance of in-line inspection tool. Ultrasonic tool are particularly sensitive to wax deposits because of the acoustic and elastic properties of the wax.

Over time, wax deposition can significantly reduce the efficiency of a pipeline system by reducing the cross sectional area of the pipe and/or by significantly increasing the pressure drop along the line. In extreme cases, wax deposition can result in the blockages of both the pipeline and associated equipment (especially in production gathering systems).

Wax compounds have also been used for external pipeline coatings for over 50 years. See coating.

Weibull distribution: A commonly used distribution for modeling extreme values. The probability density function , $n(x)$, of the Weibull distribution is given by

$$w(x) = \frac{\lambda}{\kappa} \left(\frac{x}{\lambda} \right)^{\kappa} e^{-(x/\lambda)^{\kappa}}$$

and the CDF, $W(x)$, is given by

$$W(x) = 1 - e^{-(x/\lambda)^{\kappa}}$$

where μ = location parameter of the distribution
       σ = shape parameter

Table 92 shows selected properties of the Weibull distribution. Figure 93 shows the shape of the Weibull distribution for various values of μ and σ.

Table 92. Weibull distribution properties

Parameters	λ and κ
Domain	$0 \le x < \infty$
Mean	$\lambda\Gamma\left(1+\dfrac{1}{\kappa}\right)$
Variance	$\lambda^2\Gamma\left(1+\dfrac{2}{\kappa}\right)-\left(\lambda\Gamma\left(1+\dfrac{1}{\kappa}\right)\right)^2$
Mode	$\kappa\left(\dfrac{\kappa-1}{\kappa}\right)^{1/\kappa}$

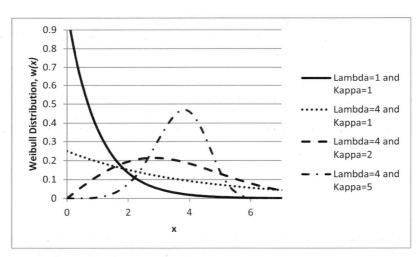

Figure 93. Plot of PDF when using Weibull distribution.

weld: The process of joining two metals together. In the context of pipe fabrication, joining the two ends of a rolled plate (i.e., a long-seam weld) to form a pipe joint or the process of joining two sections of pipe (i.e., a girth weld) to form a single section. In pipeline applications the most common type of welding is *arc welding,* which is typically used without the application of pressure where the added (molten) filler flows due to arc force and surface tension. One of the unique requirements of arc welding is shielding from atmospheric contaminants (e.g., oxygen, hydrogen, nitrogen, and water vapor) that lower the quality of the weld. The weld is laid down in several passes to accommodate typical wall thicknesses while maintaining weld quality. Welds, and weld-affected areas, can be of particular concern as changes/defects introduced during the joining process can serve as initiation points for flaws that significantly impact pipeline integrity. Welding also plays a key role in pipeline repair and maintenance because hot tapping, stopples, sleeve installation, among other techniques, all require welding. Table 93 shows the most common types of arc welding in pipeline applications.

Table 93. Common welding processes for pipeline applications [adapted from (6)]

Type	Description
Shielded Metal Arc Welding (SMAW)	• Most widely used type of arc welding – also referred to as *stick welding*. • Coalescence is achieved by heating with an electric arc between a consumable covered (or coated) electrode and the work surface. • Shielding is provided by decomposition of the electrode covering. • Filler metal is obtained from the metal core.
Gas Metal Arc Welding (GMAW)	• Formerly known as "MIG" welding. • Coalescence is achieved by the heat of an electric arc maintained between the end of a consumable electrode and the work surface. • The electrode is typically in the form of a wire that is continuously fed. • Shielding of the arc is provided by a gas (or gas mixture), which may or may not be inert.
Gas Tungsten Arc Welding (GTAW)	• Commonly known as *TIG welding*. • Coalescence is achieved by the arc and electrode method except that the tungsten electrode is not consumed. • Shielding is provided by an inert gas. • Filler metal is added externally. • This process can produce high-precision welds but requires the highest level of skill.
Flux Cored Arc Welding (FCAW)	• Coalescence by means of an arc between a continuous consumable electrode and the work surface. • Shielding is provided by flux contained within the tubular electrode. • Additional alloying elements are often contained within the core. • Additional shielding may be obtained from a gas or gas mixture.
Submerged Arc Welding (SAW)	• Coalescence is produced by the heat of an arc between a bare metal electrode and the work surface. • Arc is shielded by a blanket of granular, fusible flux. • No visible evidence of the passage of current between the electrode and the work surface. • Eliminates the sparks, spatter, and smoke ordinarily seen in other arc welding processes. • Fumes are still produced, but not in quantities generated by other processes.

weld area corrosion: A generic descriptor of corrosion found in the weld area; the term does not refer to a specific mechanism or type of corrosion.

weld bead: The filler material deposited during a single pass of the welding process. See *weld*.

weld corrosion: A generic descriptor of corrosion found in the weld area; it does not refer to a specific mechanism or type of corrosion.

weld cracking: A generic descriptor of cracking found in the weld area; it does not refer to a specific mechanism or type of cracking.

weld decay: Refers to various mechanisms (or combination of mechanisms) that lead to general degradation of a weld. Thus, the term is a generic descriptor and does not refer to a specific mechanism or type of degradation.

weld defect: An anomaly in the weld area that was introduced during the welding process. The term is a generic descriptor and does not refer to a specific type of defect. Examples of weld defects include, but are not limited to, porosity, solidification cracking, inclusions, and undercut. It should be noted that not all weld defects are an integrity concern and each must be evaluated based on the specifics of the situation and nature of the defect. A brief summary of major weld defects appears in Table 94.

weld deposition repair: This type of repair restores wall thickness by setting down weld metal directly to the area of wall loss. This method has both proponents and detractors, as shown in Table 95.

While there is evidence of "*puddle welding*" in old pipelines, this is not a prevalent technique in industry. There are some references dating back to 1958 (ICI, Battelle, British Gas) where the feasibility of weld deposition repair was established; however, until recently, in-service repair by weld deposition was not permitted for either Canadian or US gas transmission pipelines; it is now deemed acceptable under very specific situations. The technique is also permitted in pipe mills (API 5L - although it is followed by hydrostatic test) and in other industries. A guide pertaining to the practical aspects of weld deposition repair, including acceptance criteria for discontinuities detected during inspection is available through the PRCI website.

weld feature: An anomaly within a weld as detected by an in-line inspection tool. See *weld defect*.

weld material: The filler material deposited during entire the welding process. See *weld* for a more in thorough discussion of the relevant terms.

weld metal solidification cracking: See *solidification cracking*.

weld porosity: Refers to the void(s) in weld metal that result when the flux is damp or if flux cover or shielding gas is lost. The voids can be in the form of isolated spheres, clusters, or cylindrical gaps.

weld spatter: Irregular localized coating holidays adjacent to the girth weld caused by hot material from the welding process burning through the coating. The only solution is to ensure that the pipe, beyond the cutback area, is well protected during the welding process.

welding: The process of creating a weld. See *weld*.

welding passes: A layer of weld material laid down by traversing the weld area a single time. The welding process requires several passes that are often classified as indicated Figure 51.

wet film gauge: An instrument used to measure the thickness of paints, coatings, adhesives, or other covering freshly applied to the pipe body.

Table 94. Common weld defects (14)

Defect	Description
Arc strike	A localized heat-affected zone, which may have a gouge-like appearance on the surface of the pipe. See *arc strike*.
Burn-through	A localized area where the weld root bead is missing. *See burn-through*.
Cap undercut	A localized phenomenon where the cap bead is not sufficiently large and results in a groove between the weld bead and parent material. See *cap undercut*.
Concave root	A situation where the root bead has a concave profile resulting in a somewhat incomplete weld. See *excess concavity*.
Excess cap height/ penetration	A situation where the cap bead has excess material – specifically, it has a height > 3mm. See *excess convexity*.
Excess penetration	A situation where too much heat in the weld area results in too much melting of the parent material (and in extreme cases weld metal drips out from the bottom of the weld). See *excess penetration*.
Fish eyes	A situation where hydrogen causes fissures in the weld metal. See *fish eyes*.
Hydrogen cracking	A situation where localized cracking in the weld area has occurred (specifically in areas where hydrogen embrittlement has resulted in a localized loss of ductility). See *hydrogen damage*.
Inclusion	A generic term describing a situation where impurities have been incorporated into the weld metal. More specific types of inclusions include tungsten inclusion and *slag inclusion*.
Lack of fusion	A localized area where the weld metal has not fused with either the adjacent bead and/or parent material – usually a result of improper technique or insufficient heat. See *lack of fusion*.
Lack of penetration	A situation where a gap remains between the two components of the parent material due to improper technique or insufficient deposition of the weld metal. See *lack of penetration*.
Misalignment	A weld where the ends of the two pipe joints have been misaligned in some way. See *misalignment*.
Misplaced cap	A weld where the final weld pass fails to cover the previous bead (potentially resulting in excess cap height in another area of the weld). See *misplaced cap*.
Porosity	A situation where the weld contains isolated spheres or voids resulting from a problem with the flux or shielding gas. See *weld porosity*.
Root undercut	A situation where a linear defect is introduced next to the weld bead from the force of the welding arc or angle of the welding rod. See *undercut*.
Solidification crack	A crack that forms in the center of the weld bead (i.e., internally); there can be multiple causes for this. See *solidification cracking*.
Start/stop defect	A situation where improper technique results in a lack of fusion between the weld metal and parent material at the beginning or end of a weld bead.

wet gas: The definition of wet gas is not standardized across the sector; however, there is the common theme of a gaseous product that contains significant levels of liquids. **1)** Natural gas containing liquefiable hydrocarbons – natural gasoline, butane, pentane, and other light hydrocarbons that can be removed by chilling, pressure, or extraction; **2)** The percentage of methane; that is, "wet" gas contains less than 85% methane. **3)** Perhaps most relevant in the context of pipeline integrity is that "wet gas" has water in excess of 7 pounds per million cubic feet (mmcf). The presence of liquids in a gas line can lead to internal erosion (depending on specific flow conditions); further, the presence of water is a further complicating factor in that it can lead to greater concerns regarding internal corrosion.

Table 95. Advantages and disadvantages of weld deposition repair (6)

Advantages	Disadvantages
• Environmental incentives (reduces release of product and minimizes use of new pipe)	• Burn-through is a significant risk; remaining ligaments tend to be thin
• Can be used where full-encirclement sleeves cannot (e.g., bend sections and fittings)	• Resulting pipeline integrity may not be adequate (increased risk of hydrogen cracking)
• Relatively quick and inexpensive method	• Difficult to restore static strength and resistance to pressure cycles

wheel slippage: Refers to a situation where the odometer wheel on an inspection tool rotates freely due to the build-up of debris or as a result of mechanical failure. The result is that the distance indicated in the data record is shorter than the actual length of line inspected. Because slippage is not usually a consistently occurring condition, locating anomalies and other pipeline features can become more difficult – in extreme cases, a re-inspection the line may be required.

white metal blast-cleaned surface: The term implies the removal of all mill scale, rust, rust scale, paint, or foreign matter by abrasives propelled through nozzles or by centrifugal wheels. Such a surface is defined as a surface with a gray-white, uniform metallic color, slightly roughened to form a suitable anchor pattern for coatings. The surface, when viewed without magnification, shall be free of all oil, grease, dirt, visible mill scale, rust, corrosion products, oxides, paint, or any other foreign matter.

wireline cable: The tether that connects a tethered tool to its source of power and data processing and storage system. See *wireline tool*.

working electrode: In the context of a cathodic protection system, the electrode that is associated with the location of interest.

wireline tool: a class of tools that are physically connected to a cable or tether during operation. The cabling is used for various functions, including tool retrieval, power and data transmission. These tools have their origins in oilwell drilling and completion, but also have a relevant application in relatively short (typically <5 km) and small diameter (typically <12 in.) pipelines.

wrinkle: Refers to the ripples that occur on the inner radius of a pipe when the pipe is subjected to cold-bending methods. Wrinkles are not desirable and can be a significant concern from a pipeline integrity perspective. Wrinkles can also occur as a result of geotechnical movements that subjects the pipeline to unexpected stresses and strains. The term is also used to describe the *sagging* of coatings (especially *tape coatings*) that may be observed under soil loadings.

WT: See *nominal wall thickness*.

X

X-ray: The technique of obtaining a shadow image of a solid using penetrating radiation, such as x-rays or gamma-rays. The image obtained is in projection, with no details of depth within the solid. The images are recorded on film. The contrast is due to different degrees of absorption of x-rays in the specimen and depends on variations in specimen thickness, different chemical constituents, non-uniform densities, flaws, discontinuities, or scattering processes within the specimen. Access to opposite sides of the object is required. The most common application of this technique in the pipeline industry is for quality control of girth welds.

Y

Yellow Jacket®: A proprietary factory-applied extruded polyethylene (external) pipeline coating formulation although the term is often used generically to indicate the general class of coatings. See *coating*.

yield plot: The plot of the volume vs. pressure during *hydrotest*.

yield point: The point at which an increase of stress will cause it to deform plastically (i.e., permanently). Up until this point, any deformation is referred to as elastic and is not permanent; that is, once the stress is removed, the material returns to its original dimensions. Typically, the deformation of a metallic is on the order of 0.2%. See *stress-strain curve*.

yield strength: The maximum stress that a material can withstand with no (or minimal) permanent deformation. Any increase in the stress above the yield strength causes the material to exhibit a specified permanent deformation or produces a specified total elongation under load. This is a key factor in determining the required pipe wall thickness for pipeline design and can be determined from the stress-strain curve for the material. See *stress-strain curve*.

Young's modulus: An elastic constant of materials that defines the relationship between an applied elastic stress and the resulting change in strain under uniaxial loading. Young's modulus, E, is the ratio of the applied stress to the change in length of a unit of length (or change in width per unit of width):

$$E = \frac{stress}{strain}$$

where $stress$ = applied pressure (force per unit area)
  $strain$ = resulting new length/old length (alternatively, width/old width)

The Young's modulus is also related to other moduli such as

$$K = \frac{E}{2(1-2v)}$$

where K = bulk modulus
  v = Poisson's ratio

Figure 94 illustrates the relationship between stress and strain expressed by Young's modulus.

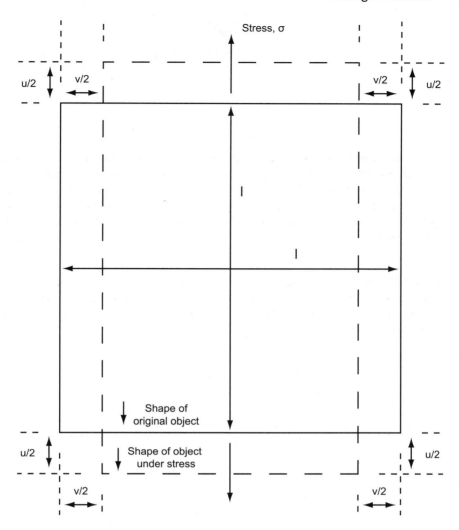

Tensile strain = u/l
Lateral strain = v/l
Young's modulus, E = stress / strain

Figure 94. Inputs for calculating of Young's modulus. (10)

Z

zero resistance ammetry: A technique that measures the voltage drop (directly proportional to the galvanic current flow) between two differing electrodes by imposing a "zero" voltage drop on the external circuit. The electrodes may be physically separate or may represent areas of dissimilar materials on a single piece of pipe. For example, the technique may be used to measure the preferential corrosion of weld metal vs. the pipe body.

zinc anodes: One of the most common types of sacrificial anodes used in galvanic cathodic protection systems; the other common material used for these types of anodes is magnesium. Zinc anodes result in better performance at low current levels when compared to magnesium anodes.

Appendices

Appendix 1.

NACE Standards Relating to Pipeline Integrity (184)

Note: It is best practice to determine and use the latest published version of any Standard or Recommended Practice. This information can be obtained directly from the organizations issuing the Standard or Recommended Practice.

Standard No	Title	Description
• NACE No. 1 • SSPC-SP 5	White metal blast cleaning	Presents requirements for white metal blast cleaning of steel surfaces by the use of abrasives. Defines the cleaning and includes sections on procedures before blast cleaning, blast-cleaning methods and operation, blast-cleaning abrasives, procedures following blast cleaning, inspection, and safety and environmental requirements.
• NACE No. 2 • SSPC-SP 10	Near-white metal blast cleaning	Presents requirements for near-white metal blast cleaning of steel surfaces by the use of abrasives. Defines the cleaning and includes sections on procedures before blast cleaning, blast-cleaning methods and operation, blast-cleaning abrasives, procedures following blast cleaning, inspection, and safety and environmental requirements.
• NACE No. 3 • SSPC-SP 6	Commercial blast cleaning	Presents requirements for commercial blast cleaning of steel surfaces by the use of abrasives. Defines commercial blast cleaning and includes sections on procedures before blast cleaning, blast-cleaning methods and operation, blast-cleaning abrasives, procedures following blast cleaning, inspection, and safety and environmental requirements.
• NACE No. 4 • SSPC-SP 7	Brush-off blast cleaning	Presents requirements for brush-off metal blast cleaning of steel surfaces by the use of abrasives. Defines brush-off blast cleaning and includes sections on procedures before blast cleaning, blast-cleaning methods and operation, blast-cleaning abrasives, procedures following blast cleaning, inspection, and safety and environmental requirements.
• NACE No. 5 • SSPC-SP 12	Surface preparation and cleaning of metals by water jetting prior to recoating	Defines four levels of visible surface cleanliness (WJ-1 to WJ-4), three levels of flash rusting, and three levels of invisible cleanliness (NV-1 to NV-3) for surface preparation using water jetting. Key words: cleaning, coatings, high-pressure water jetting, steels.

Standard No	Title	Description
NACE RP0102	In-line inspection of pipelines	The standard relates specifically to planning, organizing, and executing the in-line inspection of a pipeline section. The guidance document is targeted to personnel involved with in-line inspection within an operations company. As such, the standard includes references to several key steps in undertaking an in-line inspection including, but not limited to, determining piggability, cleaning/gauging requirements, contracting, inspection procedures, tool tracking along with some reference to data analysis.
NACE RP0104	The use of coupons for cathodic protection monitoring applications	Addresses applications for cathodic protection (CP) coupons attached to buried pipelines to determine the level of corrosion protection provided by a CP system. Appendixes cover coupons attached to other structures, such as underground storage tanks (USTs), aboveground (on-grade) storage tank bottoms, and steel-in-reinforced concrete structures. CP coupons may also be used to evaluate compliance with CP criteria, including considering the IR drop.
NACE RP0105	Epoxy coatings for external repair, rehabilitation, and weld joints on buried steel pipelines	Presents guidelines for establishing minimum requirements to ensure proper material selection, application, and inspection of pipeline liquid-epoxy coatings used for the repair and rehabilitation of previously coated pipelines and for coating field joints on the external surfaces of pipe.
NACE RP0208	Internal corrosion direct assessment methodology for liquid petroleum pipelines	Describes the basis of the liquid petroleum internal corrosion direct assessment (LP-ICDA) method and its four steps: pre-assessment, indirect assessment, direct examination, and post assessment. With the LP-ICDA approach, assessments can be performed on pipe segments for which alternative methods (e.g., in-line inspection, hydrostatic testing, etc.) may not be practical. This methodology may be incorporated into corrosion integrity and risk management plans.
NACE RP0274	High-voltage electrical inspection of pipeline coatings	Provides information on high-voltage electrical inspection of pipeline coatings. Gives guidelines on testing voltages, grounding, exploring electrodes, speed of travel, voltage measurements, surface condition, and care of equipment.
NACE RP0303	Field-applied heat-shrinkable sleeves for pipelines: application, performance, and quality control	Provides current technology and industry practices for the use of field-applied heat-shrinkable sleeve systems. Included are methods for qualifying and controlling the quality of a heat-shrinkable sleeve, guidelines for proper application, and inspection and repair techniques to ensure its long-term performance. Key words: protective coatings, field joints, heat-shrinkable sleeves, pipelines.

Standard No	Title	Description
NACE RP0304	Design, installation, and operation of thermoplastic liners for oilfield pipelines	Defines the process necessary to design, install, and operate a thermoplastic-lined oilfield pipeline. Intended to set a standard for performance within the scope of procedures and instructions already developed by installers and operators. Key words: internal pipe coatings, liners, linings, pipelines, protective coatings, thermoplastic.
NACE RP0375	Field-applied underground wax coating systems for underground pipelines	Provides material requirements, surface preparation recommendations, and application techniques for the successful use of hot- or cold-applied wax and component wrappers and wax tape coating systems for the protection of pipes, fittings, and valves. Lists recommend physical requirements for different types of coating systems and the corresponding ASTM standard test methods.
NACE RP0394	Application, performance, and quality control of plant-applied, fusion-bonded epoxy external pipe coating	Presents guidelines for establishing minimum requirements to ensure proper application and performance of plant-applied, fusion-bonded epoxy coatings to the external surfaces of pipe. Describes methods for qualifying and controlling the quality of fusion-bonded epoxy pipe coatings, provides guidelines for their proper application, and identifies inspection and repair techniques to ensure their long-term performance. Appendixes present test methods to be used in conjunction with the criteria outlined in the body of the standard.
NACE RP0399	Plant-applied, external coal tar enamel pipe coating systems: application, performance, and quality control	Presents guidelines for establishing minimum requirements to ensure proper performance of coal tar enamel coating systems. Recommended physical property requirements and surface preparation, application, inspection, and repair practices are described.
NACE RP0402	Field-applied fusion-bonded epoxy (FBE) pipe coating systems for girth weld joints: application, performance, and quality control	Provides the most current technology and industry practices for the use of field-applied fusion-bonded epoxy (FBE) external pipe-coating systems for girth weld joints. This standard is intended for use by corrosion-control personnel, design engineers, project managers, purchasing personnel, and construction engineers and managers. It is applicable to underground steel pipelines in the oil and gas gathering, distribution, and transmission industries.
NACE RP0497	Field corrosion evaluation using metallic test specimens	Describes how field corrosion evaluation using metallic test specimens is conducted, what types of corrosion information may be obtained, and how test racks and specimens are designed. A summary of critical data that must be recorded is provided. Guidelines for interpreting and reporting test results are also discussed.

Standard No	Title	Description
NACE RP0602	Field-applied coal tar enamel pipe coating systems: application, performance, and quality control	Provides the most current technology and industry practices for the use of field-applied coal tar enamel external pipe coating systems. This standard is intended for use by corrosion-control personnel, design engineers, project managers, purchasing personnel, and construction engineers and managers. It is applicable to underground steel pipelines in the oil and gas gathering, distribution, and transmission industries.
NACE RP0607	Petroleum and natural gas industries – cathodic protection of pipeline	This ANSI/NACE standard is a modified adoption of ISO Standard 15589-2. It specifies requirements and gives recommendations for the pre-installation surveys, design, materials, equipment, fabrication, installation, commissioning, operation, inspection, and maintenance of cathodic protection systems for offshore pipelines for the petroleum and natural gas industries, as defined in ISO 13623. This standard is applicable to carbon and stainless steel pipelines in offshore service, retrofits, modifications, and repairs made to existing pipeline systems. It applies to all types of seawater and seabed environments encountered in submerged conditions and on risers up to mean water level.

Standard No	Title	Description
NACE SP0106	Control of internal corrosion in steel pipelines and piping systems	Describes procedures and practices for achieving effective control of internal corrosion in steel pipe and piping systems in crude oil, refined products, and gas service. Because of the complex nature and interaction between constituents that are found in gas and/or liquid, certain combinations of these impurities being transported in the pipeline may affect whether a corrosive condition exists. Identification of corrosive gas and/or liquid in a pipeline can only be achieved by analysis of operating conditions, impurity content, physical monitoring, and/or other considerations.
NACE SP0169	Control of external corrosion on underground or submerged metallic piping systems	The cathodic protection criteria in this standard for achieving effective control of external corrosion on buried or submerged metallic piping systems are also applicable to other buried metallic structures. The standard includes information on determining the need for corrosion control; piping system design; coatings; cathodic protection criteria and design; installation of cathodic protection systems; and control of interference currents. The cost of corrosion control is also addressed in the appendices.
NACE SP0177	Mitigation of alternating current and lightning effects:	Presents guidelines and procedures for use during design, construction, operation, and maintenance of metallic structures and corrosion-control systems used to mitigate the effect of lightning and overhead alternating current (AC) power transmission systems.
NACE SP0185	Extruded polyolefin resin coating systems with soft adhesives for underground or submerged pipe	Details materials and methods of application for two types of polyolefin resin coating systems extruded over soft adhesives on pipe for underground or submerged service. The standard addresses surface preparation, application methods, electrical inspection, pipe-handling techniques, and coating system repair methods. The two types of coating systems are polyolefin resin that is crosshead-extruded on the pipe as a seamless coating over a hot-applied mastic adhesive and polyolefin resin that is extruded spirally around the pipe to fuse and form as a seamless coating over an extruded butyl-rubber adhesive.
NACE SP0200	Steel-cased pipeline practices	Details acceptable practices for the design, fabrication, installation, and maintenance of steel-cased metallic pipelines. It is intended for use by personnel in the pipeline industry.

Title	Description
Internal corrosion direct assessment methodology for pipelines carrying normally dry natural gas	Formalizes the process of internal corrosion direct assessment (ICDA) for pipelines carrying normally dry natural gas that can be used to help ensure pipeline integrity. The basis of DG-ICDA is a detailed examination of locations along a pipeline where water would first accumulate to provide information about the downstream condition of the pipeline. If the locations along a length of pipe most likely to accumulate water have not corroded, other downstream locations less likely to accumulate water may be considered free from corrosion.
NACE SP0207 — Performing close-interval potential surveys and dc surface potential gradient surveys on buried or submerged metallic pipelines	Presents procedures for performing close-interval DC pipe-to-electrolyte potential surveys on buried or submerged metallic pipelines. This standard addresses the potential survey component of hybrid survey techniques such as trailing-wire DCVG or intensive measurement surveys, but does not address other surveys such as cell-to-cell techniques used to evaluate the direction of current or the effectiveness of the coating. This standard is intended for use by corrosion control personnel involved with operating pipelines, contractors performing close-interval surveys, corrosion professionals interpreting close-interval survey data, and regulatory agencies.
NACE SP0286 — Electrical isolation of cathodically protected pipelines	Fully details the requirements necessary to ensure adequate isolation of cathodically protected pipelines, especially those with high-quality dielectric coatings. The standard was developed as a supplement to RP0169, RP0675, and RP0177. It includes sections on the need for electrical isolation, methods of electrical isolation, comparison of devices available for pipeline isolation, and equipment specification and installation, as well as field testing and maintenance.
NACE SP0487 — Considerations in the selection and evaluation of rust preventatives and vapor corrosion inhibitors for interim corrosion protection	Discusses considerations in selection and performance criteria of interim coatings. Quality control criteria are listed to enable the manufacturer and user to select appropriate test procedures to maintain prescribed standards. The standard is intended to assist the new buyer or user as well as the experienced user of interim coatings in the proper selection and evaluation of these coatings.

Standard No	Title	Description
NACE SP0490	Holiday detection of fusion-bonded epoxy external pipeline coatings of 250 to 760 μm (10 to 30 mil)	Presents recommended techniques in the operation of holiday detector equipment currently used on fusion-bonded epoxy (FBE) pipeline coatings following shop application of the coating and prior to on-site installation of the coated pipeline. It also presents recommended voltages for various coating thicknesses. This standard is intended to serve the needs of pipeline owners, coating applicators, coating inspectors, and other interested parties in the electrical inspection of FBE pipe coatings.
NACE SP0492	Metallurgical and inspection requirements for offshore pipeline bracelet anodes	Sets minimum physical quality and inspection standards for cast sacrificial anodes for offshore pipeline applications. The standard is applicable to typical half-shell or segmented bracelet-type anodes and is not intended to apply to platform, hull, tank, or extruded-type anodes. The section on physical requirements includes information on samples for chemical analysis; anode identification, weight, dimensions, and straightness; insert dimensions and position; insert material; fabrication of inserts by welding; insert surface preparation; surface irregularities on the anode casting; cracks in cast anodic materials; defects; and more.
NACE SP0502	Pipeline external corrosion direct assessment methodology	Covers the NACE external corrosion direct assessment (ECDA) process – a process of assessing and reducing the impact of external corrosion on pipeline integrity. ECDA is a continuous improvement process providing the advantages of locating areas where defects can form in the future, not just areas where defects have already formed, thereby helping to prevent future external corrosion damage. This standard covers the four components of ECDA: Pre-Assessment, Indirect Inspections, Direct Examinations, and Post Assessment.

Standard No	Title	Description
NACE SP0507	External corrosion direct assessment (ECDA) integrity data exchange (IDX) format.	Joint NACE/PODS standard. The objective of this standard practice is the development of a new external corrosion direct assessment (ECDA) data interchange data structure in order to enable electronic integration of data and standardize reporting of ECDA data within the pipeline industry to allow transfer between different software packages or computer systems. This is expected to minimize difficulty in using various programs to analyze or graph data and allow for comparison of data gathered for a given pipeline segment at different times, regardless of the software system used to collect it. The format outlined is the commonly used American Standard Code for Information Interchange (ASCII) comma delimited text file, which is adaptable to all data processing systems. This standard is expected to serve as a template for future internal corrosion direct assessment (ICDA) and stress-corrosion cracking direct assessment (SCCDA) data interchange standards.
NACE SP0572	Design, installation, operation and maintenance of impressed current deep anode beds.	Presents procedures and practices for design, installation, operation, and maintenance of deep groundbeds used for control of external corrosion of underground or submerged metallic structures by impressed current cathodic protection. Key words: anodes, deep groundbeds.

Standard No	Title	Description
NACE Technical Committee Report 10A392	Effectiveness of cathodic protection	This technical committee report was prepared as an information guide for external corrosion control of thermally insulated underground metallic surfaces and considerations of the effectiveness of cathodic protection. Although pipelines are the primary focus of this report, the principles discussed would be applicable when a thermal insulating material has been applied on or in the immediate proximity of an underground metallic surface (6).
NACE Technical Committee Report 10D199	Coatings for the repair and rehabilitation of the external coatings of buried steel pipelines	Provides detailed, generic product information on the 14 types of pipeline coatings most often used for repair and rehabilitation projects, and general guidance on coating selection. Information on surface preparation, application methods, and inspection is also included.

Standard No	Title	Description
NACE Technical Committee Report 1E100	Engineering symbols related to cathodic protection	Illustrates commonly used engineering symbols and graphic representations that are related to cathodic protection used in corrosion control. Because cathodic protection symbols have no current uniform designations, this state-of-the-art report is intended to provide a straightforward reference of commonly used symbols for engineers, designers, contractors, and other corrosion control personnel who use cathodic protection. This report is designed to assist in avoiding confusion among design, construction, and other associated corrosion activities.
NACE Technical Committee Report 05107	Report on corrosion probes in soil or concrete	Electrical resistance (ER) corrosion measurement probes, typically used for internal corrosion monitoring, have been adapted for use in soil-side applications. They are commonly referred to as soil corrosion probes (SCP) when used in soil-side applications. SCP are often used to determine the effectiveness of corrosion protection applied to a variety of structures such as buried pipelines, underground and aboveground storage tanks, and reinforced concrete. This report mainly addresses applications for ER SCP attached to buried pipelines. It also includes information on linear polarization resistance (LPR) probes used for assessing soil corrosiveness and on the use of corrosion probes for reinforced concrete structures. This report is intended for use by corrosion control personnel responsible for monitoring the condition of buried structures or those in contact with an electrolyte.
NACE Technical Committee Report 05101	State-of-the-art survey on corrosion of steel pilings in soils	Provides information on the design, maintenance, and rehabilitation of structures with steel pile foundations. The report contains information obtained from a survey of the open literature on corrosion of steel pilings
NACE Technical Committee Report 30105	Electrical isolation/continuity and coating issues for offshore pipeline cathodic protection systems	Provides owners, engineers, contractors, and operators with information on electrical isolation/continuity issues and coating issues to consider when designing and operating offshore pipeline cathodic protection (CP) systems. The detailed specifications for application of pipeline coatings are beyond the scope of this report, though it gives references to some specifications that are used. This report is a support document for the pipeline CP design standards.

Standard No	Title	Description
NACE Technical Committee Report 35100	In-line non-destructive inspection of pipelines	Analyzes available and emerging technologies in the field of in-line pipeline inspection tools and reviews their status with respect to characteristics, performance, range of application, and limitations. It is intended as a practical reference for both new and experienced users of in-line inspection technology. Assists in the provision of an understanding of the practical aspects of using the tools, highlighting the implications, and helping assess the benefits.
NACE Technical Committee Report 35101	NACE technical committee report 35101 plastic liners for oilfield pipelines	Provides an overview of thermoplastic liners used in oilfield pipelines, and reflects current practices. This report is intended to assist those who are considering the use of liners, but have only limited access to resources with knowledge of the terminology, techniques, and applications of liners in the oil field.
NACE Technical Committee Report 35103	External stress corrosion cracking of underground pipelines	Provides useful information on stress corrosion cracking (SCC) for engineers, designers, consultants, and others involved in the design, maintenance, and rehabilitation of underground petroleum (including gas, crude oil, and refinery products) pipelines. This technical committee report contains information obtained from a survey of the open literature on the subject.
NACE Technical Committee Report 35108	One hundred millivolt (mv) cathodic polarization criterion	This report discusses the theoretical basis for the 100 millivolt (mV) cathodic polarization criterion, the effects of other factors such as temperature, mill scale, moisture, and anaerobic bacteria, measurement of the polarization, and the applicability of the criterion in situations, such as areas susceptible to stress-corrosion cracking, mixed-metal systems, and areas susceptible to stray currents. It also includes the results of an industry questionnaire on the use of the 100 mV polarization criterion and opinions on its effectiveness.

Standard No	Title	Description
NACE Technical Committee Report 35201	Technical report on the application and interpretation of data from external coupons used in the evaluation of cathodically protected metallic structures	NACE Technical Committee Report 35201 Technical Report on the Application and Interpretation of Data from External Coupons Used in the Evaluation of Cathodically Protected Metallic Structures: Presents state-of-the-art technical information on the applications, designs, and installation of CP coupons to determine the level of corrosion protection provided by a cathodic protection (CP) system to a buried structure. Provides an overview of equipment commonly used to monitor CP coupons and interpretations of monitoring results. This technical committee report is intended to assist corrosion personnel who use CP coupon data to evaluate the corrosion and cathodic protection levels of their structures better.
NACE Technical Committee Report 6A100	Coupons used in the evaluation of cathodically protected metallic structures	Information on two protective systems: coatings and cathodic protection. With specific information on the strengths and weaknesses of each protective system and explains how a greater level of corrosion protection can be achieved on buried or immersed substrates if both systems are used together.
NACE Technical Committee Report 6G198	Wet abrasive blast cleaning:	Describes the procedures, equipment, and materials used for air/water/abrasive, water/abrasive, and pressurized water/abrasive blasting. Information on inhibitors, equipment maintenance, and safety guidelines is also included.
NACE TM0102	Measurement of protective coating electrical conductance on underground pipelines	Presents guidelines and procedures for use primarily by corrosion control personnel in the pipeline industry to determine the general condition of a pipeline coating. These techniques are used to measure the coating conductance (inverse of coating resistance) on sections of underground pipelines. This test method applies only to pipe coated with dielectric coatings. Key words: coating conductance, corrosion testing, measurement.

Standard No	Title	Description
NACE TM0106-2006	Detection, testing, and evaluation of microbiologically influenced corrosion on external surfaces of buried pipelines	This standard is for ferrous-based metal pipeline facilities and describes types of microorganisms, mechanisms by which MIC occurs, methods of testing for the presence of bacteria, research results, and interpretation of testing results. Discussed in this standard are the technical aspects of MIC, field equipment and testing procedures, and media and techniques that can be used for testing. This standard is intended for use by pipeline operators, pipeline service providers, government agencies, and any other persons or companies involved in planning or managing pipeline integrity. Key words: microbiologically influenced corrosion, pipelines, sampling and test procedures.
NACE TM0172	Determining corrosive properties of cargoes in petroleum product pipelines	Provides a uniform method of testing the corrosive properties of petroleum product pipeline cargoes. This standard provides guidelines for performing the test method described in ASTM D 665, modified so that it is applicable to gasoline and other petroleum products, and so that it permits analysis within a single working day. This short test is particularly applicable to a batch control procedure because of the need for prompt release of cargoes and because time is limited during the working day.
NACE TM0194	Field monitoring of bacterial growth in oil and gas systems	Standard based on a report from the former Corrosion Engineering Association (CEA) of the United Kingdom. Intended for use by technical field and service personnel. Describes field methods for estimating bacterial populations, including sessile bacterial populations, commonly found in oilfield systems. Sampling methods are emphasized and media for enumerating common oilfield bacteria are described. Key words: bacteria, corrosion testing, field monitoring, oilfield production equipment.

Standard No	Title	Description
NACE TM0284	Evaluation of pipeline and pressure vessel steels for resistance to hydrogen-induced cracking	Provides a standard set of test conditions for consistent evaluation of pipeline and pressure vessel steels and comparing test results from different laboratories pertaining to the results of the absorption of hydrogen generated by corrosion of steel in wet H_2S. Describes two test solutions, Solution A and Solution B, and includes special procedures for testing small-diameter, thin-wall, electric-resistance welded and seamless line pipe. Test is intended to evaluate resistance to hydrogen-induced (stepwise) cracking only, and not other adverse effects of sour environments such as sulfide stress cracking, pitting, or weight loss from corrosion. Complements NACE Standard MR0175. Key words: steel pipelines, sulfide stress cracking.
NACE TM0497	Measuring techniques related to criteria for cathodic protection on underground or submerged metallic piping systems	Provides testing procedures to comply with the requirements of a criterion at a test site on a buried or submerged steel, cast iron, copper, or aluminum pipeline. Contains instrumentation and general measurement guidelines. Includes methods for voltage drop considerations when making pipe-to-electrolyte potential measurements and provides guidance to prevent incorrect data from being collected and used.

Appendix 2.

API and ASME Standards Relating to Pipeline Integrity

Note: It is best practice to determine and use the latest published version of any Standard or Recommended Practice. This information can be obtained directly from the organizations issuing the Standard or Recommended Practice.

Standard No	Title	Description
API 80	Guidelines for the definition of onshore gas gathering	Published by the American Petroleum Institute and developed by an industry coalition that included representatives from more than 20 petroleum industry associations., it provides a functional description of onshore gas gathering pipelines for the sole purpose of providing users with a practical guide for determining the application of the definition of gas gathering in the federal Gas Pipeline Safety Standards, 49 *CFR* Part 192, and state programs implementing these standards.
API RP 1102	Steel pipelines crossing railroads and highways	Published by the American Petroleum Institute, this recommended practice gives primary emphasis to provisions for public safety. It covers the design, installation, inspection, and testing required to ensure safe crossings of steel pipelines under railroads and highways. The provisions apply to the design and construction of welded steel pipelines under railroads and highways. The provisions of this practice are formulated to protect the facility crossed by the pipeline, as well as to provide adequate design for safe installation and operation of the pipeline. The provisions should be applicable to the construction of pipelines crossing under railroads and highways and to the adjustment of existing pipelines crossed by railroad or highway construction. This practice should not be applied retroactively, nor to pipelines under contract for construction on or prior to the effective date of this edition, nor to directionally drilled crossings or to pipelines installed in utility tunnels.

Standard No	Title	Description
API 1104	Welding of pipelines and related facilities	A standard published by the American Petroleum Institute that covers the gas and arc welding of butt, fillet, and socket welds in carbon and low-alloy steel piping used in the compression, pumping, and transmission of crude petroleum, petroleum products, fuel gases, carbon dioxide, nitrogen and, where applicable, covers welding on distribution systems. It applies to both new construction and in-service welding. The welding may be done by a shielded metal-arc welding, submerged arc welding, gas tungsten-arc welding, gas metal-arc welding, flux-cored arc welding, plasma arc welding, oxyacetylene welding, or flash butt welding process or by a combination of these processes using a manual, semiautomatic, mechanized, or automatic welding technique or a combination of these techniques. The welds may be produced by position or roll welding or by a combination of position and roll welding.
		This standard also covers the procedures for radiographic, magnetic particle, liquid penetrant, and ultrasonic testing, as well as the acceptance standards to be applied to production welds tested to destruction or inspected by radiographic, magnetic particle, liquid penetrant, ultrasonic, and visual testing methods.
		It is intended that all work performed in accordance with this standard shall meet or exceed the requirements of this standard.

Standard No	Title	Description
API RP 1109	Marking liquid petroleum pipeline facilities	Published by the American Petroleum Institute, it addresses the permanent marking of liquid petroleum pipeline transportation facilities. It covers the design, message, installation, placement, inspection, and maintenance of markers and signs on pipeline facilities located onshore and at inland waterway crossings. Markers and signs indicate the presence of a pipeline facility and warn of the potential hazards associated with its presence and operation. The markers and signs may contain information to be used by the public when reporting emergencies and seeking assistance in determining the precise location of a buried pipeline.
		The provisions of this recommended practice cover the minimum marker and sign requirements for liquid petroleum pipeline facilities. Alternative markers, which are recommended for some locations under certain circumstances, are also discussed. The pipeline operator is responsible for determining the extent of pipeline marking. Consideration should be given to the consequences of pipeline failure or damage; hazardous characteristics of the commodity being transported; and the pipeline's proximity to industrial, commercial, residential, and environmentally sensitive areas. The pipeline marking programs are also integral parts of the pipeline operator's maintenance and emergency plans.
		This recommended practice is not intended to be applied retroactively. Its recommendations are for new construction and for normal marker maintenance programs subsequent to the effective date of current edition.
API RP 1110	Pressure testing of steel pipelines for the transportation of gas, petroleum gas, hazardous liquids, highly volatile liquids or carbon dioxide	A recommended practice standard published by the American Petroleum Institute that provides guidelines for pressure testing steel pipelines for the transportation of gas, petroleum gas, hazardous liquids, highly volatile liquids or carbon dioxide. The document provides guidance so that: • Pipeline can select a pressure tests suitable for the conditions under which the test will be conducted. This includes, but is not limited to, pipeline material characteristics, pipeline operating conditions and various types of anomalies or other risk factors that may be present. • Pressure tests are planned to meet overall objectives of the pressure test. • Site-specific procedures are developed and followed and followed during all phases of the pressure testing process. • Pressure tests consider personal safety and environmental impacts. • Pressure tests are implemented by qualified personnel. • Pressure tests are conducted to meet stated acceptance criteria and pressure test objectives. • Pressure test records are developed, completed, and retained for useful life of facility.

Standard No	Title	Description
API RP 1111	Design, construction, operation, and maintenance of offshore hydrocarbon pipeline and risers	A recommended practice published by the American Petroleum Institute that sets out criteria for the design, construction, testing, operation and maintenance of offshore steel pipelines used in the production, production support, or transportation of hydrocarbons from the outlet flange of a production facility.
		The criteria also apply to transportation piping facilities located on production platforms after separation and treatment, including meter facilities, gas compression facilities, liquid pumps, and associated piping and appurtenances.
		Limit State Design has been incorporated into the document to provide a uniform factor of safety with respect to rupture or burst failure as the primary design condition independent of the pipe diameter, wall thickness, and grade.
		The design, construction, inspection, and testing provisions of this document may not apply to offshore hydrocarbon pipelines designed or installed before this latest revision was issued. The operation and maintenance provisions of the document are suitable for application to existing facilities. Design and construction practices other than those set forth herein may be employed when supported by adequate technical justification, including model or proof testing of involved components or procedures as appropriate. Nothing identified in this document should be considered as a fixed rule for application without regard to sound engineering judgment.
API RP 1113	Developing a pipeline supervisory control center	A recommended practice published by the American Petroleum Institute that focuses on the design aspects that may be considered appropriate for developing or revamping a control center. A pipeline supervisory control center is a facility where the function of centralized monitoring and controlling of a pipeline system occurs. This document is not all-inclusive. It is intended to cover best practices and provide guidelines for developing a control center only. It does not dictate operational control philosophy or overall SCADA system functionality.
		This document is intended to apply to control centers for liquids pipelines; however, many of the considerations may also apply to gas control center design.

Standard No	Title	Description
API RP 1114	Design of solution-mined underground storage facilities	A recommended practice published by the American Petroleum Institute that provides basic guidance on the design and development of new solution-mined underground storage facilities. It is based on the accumulated knowledge and experience of geologists, engineers, and other personnel in the petroleum industry. Users of this guide are reminded that no publication of this type can be complete nor can any written document be substituted for qualified, site-specific engineering analysis. All aspects of solution-mined underground storage are covered, including selecting an appropriate site, physically developing the cavern, and testing and commissioning the cavern. Additionally, a section on plug and abandonment practices is included. This recommended practice does not apply to caverns used for waste disposal purposes. See API Recommend Practice 1115 for guidance in the operation of solution-mined underground storage facilities.
API RP 1115	Operation of solution-mined underground storage facilities	A recommended practice published by the American Petroleum Institute that provides basic guidance on the operation of solution-mined underground hydrocarbon liquid or liquefied petroleum gas storage facilities. This document is intended for first-time cavern engineers or supervisors, but would also be valuable to those people experienced in cavern operations. This recommended practice is based on the accumulated knowledge and experience of geologists, engineers, and other personnel in the petroleum industry. All aspects of solution-mined underground storage operation, including cavern hydraulics, brine facilities, wellhead and hanging strings, and cavern testing are covered. Users of this guide are reminded that no publication of this type can be complete, nor can any written document be substituted for effective site-specific operating procedures. This recommended practice does not apply to caverns used for natural gas storage, waste disposal purposes, caverns which are mechanically mined, depleted petroleum reserve cavities, or other underground storage systems which are not solution-mined.

Standard No	Title	Description
API RP 1117	Recommended practice for movement in in-service pipelines	A recommended practice published by the American Petroleum Institute that covers the design, execution, inspection, and safety of a pipeline-lowering or other movement operation conducted while the pipeline is in service. (In this document, the terms "lowering" and "movement" can be used interchangeably.) This recommended practice presents general guidelines for conducting a pipeline-movement operation without taking the pipeline out of service. It also presents equations for estimating the induced stresses. To promote the safety of the movement operation, it describes stress limits and procedures. Additionally, it outlines recommendations to protect the pipeline against damage. The practicality and safety of trench types, support systems, and lowering or other methods are considered. Inspection procedures and limitations are presented.
API RP 1130	Computational pipeline monitoring for liquids pipelines	A recommended practice published by the American Petroleum Institute that focuses on the design, implementation, testing, and operation of CPM systems that use an algorithmic approach to detect hydraulic anomalies in pipeline operating parameters. The primary purpose of these systems is to provide tools that assist pipeline controllers in detecting commodity releases that are within the sensitivity of the algorithm. It is intended that the CPM system would provide an alarm and display other related data to the pipeline controllers to aid in decision-making. The pipeline controllers would undertake an immediate investigation, confirm the reason for the alarm, and initiate an operational response to the hydraulic anomaly when it represents an irregular operating condition or abnormal operating condition or a commodity release. The purpose of this recommended practice is to assist the pipeline operator in identifying issues relevant to the selection, implementation, testing, and operation of a CPM system.

Standard No	Title	Description
API 1132	Effects of oxygenated fuels and reformulated diesel fuels on elastomers and polymers in pipeline/terminal components	With the passage of the Clean Air Act, pipelines are required to transport reformulated oxygenated products containing ethers and alcohols. In some cases pipelines are shipping neat oxygenates that may have effects on pipeline components. API surveyed the pipeline/terminal industry to determine methods for handling these products, proper selection of materials and product compatibilities. The objective of this project was to develop a document that consolidates the industry's experience with oxygenated fuels and reformulated diesel fuels and their effects on elastomers and other pipeline and storage components. The scientific objectives of this research were to identify elastomers or polymers used for different components used in pipelines now and prior to the transport or storage of oxygenated fuels; the cause of failure of components; the problems associated with oxygenated fuels on elastomer and polymer components; changes that have been made in the use of elastomer and polymer components; usefulness of published methodology for evaluating the impact of operating service variables on the service life of elastomer and/or polymer pipeline components; and problems encountered in the pipeline transportation and storage of reformulated diesel fuels.
API RP 1133	Guidelines for onshore hydrocarbon pipelines affecting high-consequence floodplains	Published by the American Petroleum Institute, this document sets out criteria for the design, construction, operation, maintenance, and abandonment of onshore pipelines that could affect high-consequence floodplains and associated commercially navigable waterways. It applies only to steel pipelines that transport gas, hazardous liquids, alcohols, or carbon dioxide. The design, construction, inspection and testing provisions of this document should not apply to pipelines that were designed or installed prior to the latest revision of this publication. The operation and maintenance provisions of this document should apply to existing facilities. The contents in this document should not be considered a fixed rule for application without regard to sound engineering judgment.

Standard No	Title	Description
API 1149	Pipeline variable uncertainties and their effects on leak detectability	A study published by the American Petroleum Institute that quantifies the effects of variables on leak detection using common software-based leak detection methods. This study provides a data base and a step-by-step methodology to evaluate leak detection potential of a given pipeline with specified instrumentation and SCADA capabilities. Incremental improvement of leak detectability resulting from upgrading individual variables can also be determined.
		The results from this study enable users (i.e., pipeline companies) to determine the achievable level of leak detection for a specific pipeline with a specified set of instrumentation and SCADA system. The results also help users to understand the sensitivity of leak detectability with respect to the variables involved. This information is useful in several ways: investigating the feasibility of leak detection systems, justifying and prioritizing changes to instrumentation and SCADA systems, configuring pipeline and measurement stations, and aiding leak detection operations.
		Three general types of software-based leak detection methods are addressed in this study: mass balance, mass balance with line flow correction, and transient flow analysis. The leak detection potential of these methods are discussed based on hydraulics to the extent possible. The liquids considered are crude oils and refined petroleum products such as gasoline, jet fuel, and fuel oil.
		The pipeline configuration considered is a pipe segment with pressure, temperature, and volumetric flow measurements at each end. During steady-state flow, this configuration applies to pipelines with booster pumping stations where rates of flow are measured only at the inlet and the outlet of the entire system. All variables affecting leak detection are listed. General relationships between the variable uncertainties and leak detection potential are analyzed. The methodology are described and verified with field tests. The variables are ranked according to their importance to leak detectability. A step-by-step method and a data base are established to enable simple hand calculations for establishing leak detectability based on mass balance. The method and the data base are verified with field data. The rationale and the procedure to establish leak detectability using mass balance with line pack correction and transient flow simulations are given and illustrated with examples and field trial results.

Standard No	Title	Description
API 1156 & 1156-A	Effects of smooth and rock dents on liquid petroleum pipelines, Phase I and Phase II	Documents published by the American Petroleum Institute that present the findings of a project sponsored by the API to determine the effects of smooth dents and rock dents on the integrity of liquid petroleum pipelines to avoid unnecessary repair or replacement of pipelines affected by dents, if they do not constitute a threat to pipeline serviceability. The addendum to the report presents a description of work that was done after the completion of Phase I. Additional work has been done to address issues confronted but not resolved in the first phase of the work, and to address new issues raised by the first-phase work.
API 1157	Hydrostatic test water treatment and disposal options for liquid pipeline systems	Published by the American Petroleum Institute, this report presents the results of a research study to define acceptable and cost effective hydrostatic test water treatment and disposal methods that enable compliance with DOT requirements for testing of liquid pipelines, while meeting regulatory agency permitting requirements for disposal and/or discharge. This study was conducted from February to December 1997 and involved data provided by 15 pipeline companies (representing about 45 % of the national pipeline system mileage.) The primary results and conclusions of this study found that activated carbon adsorption was the single most frequently used treatment technology for existing pipelines. Activated carbon adsorption appeared to be a viable option for the smaller projects of 100,000 gallons or less, but may have significant limitations for larger projects due to flow volume limitations and logistical considerations for equipment. Alternate water treatment technologies, beyond activated carbon adsorption, hay bales and air stripping, were also evaluated for cost and practicality. These alternatives included dissolved air flotation and ultraviolet light oxidation. Neither of these options proved viable for use in the pipeline industry due to cost, performance, or practicality of implementation. Other test water management options that are potentially available to the liquid pipeline industry were also identified. Inadequate data or cost information was available to evaluate these options but they may be valuable options to consider depending on the specific circumstances of the discharge. Compliance with permit discharge conditions was reported for 84 tests. Of the 329 permit conditions contained in these 84 tests, 327 (99+%) demonstrated compliance.

Standard No	Title	Description
API 1158	Analysis of DOT reportable incidents for hazardous liquid pipelines, 1986 through 1996	Published by the American Petroleum Institute, this document presents an analysis of incidents reportable to the U.S. Department of Transportation on about 160,000 miles of liquid petroleum pipelines in the U.S. during the 11-year period from 1986 to 1996. The analyses presented herein represent work conducted by the U.S. Department of Transportation's, Office of Pipeline Safety and the operators of liquid petroleum pipelines through the American Petroleum Institute to better understand the causes and consequences of incidents, to monitor trends that may indicate the need for action, to use the data to identify potential risks and where risk management would be most productive, and to identify areas for potential improvement in the data collecting process. This document includes information on general trends of the incidents, trends based on attributes, analysis of incidents by cause, and a data disk containing the incident data for the 11-year period.
API 1160	Managing system integrity for hazardous liquid pipelines	Published by the American Petroleum Institute (API), it outlines a process that an operator of a pipeline system can use to assess risks and make decisions about risks in operating a hazardous liquid pipeline to reduce the number of incidents and the adverse effects of errors and incidents. An integrity management program provides a means to improve the safety of pipeline systems and to allocate operator resources effectively to identify and analyze actual and potential precursor events that can result in pipeline incidents; examine the likelihood and potential severity of pipeline incidents; provide a comprehensive and integrated means for examining and comparing the spectrum of risks and risk reduction activities available; provide a structured, easily communicated means for selecting and implementing risk reduction activities; and establish and track system performance with the goal of improving that performance. This standard is intended for use by individuals and teams charged with planning, implementing, and improving a pipeline integrity management program. Typically a team would include engineers, operating personnel, and technicians or specialists with specific experience or expertise (corrosion, in -line inspection, right-of-way patrolling, etc.). Users of this standard should be familiar with the pipeline safety regulations (Title 49 CFR Part 195), including the requirements for pipeline operators to have a written pipeline integrity program, and to conduct a baseline assessment and periodic reassessments of pipeline management integrity.

Standard No	Title	Description
API RP 1162	Public awareness programs for pipeline operators	Published by the American Petroleum Institute, it provides guidance for operators of petroleum liquids and natural gas pipelines to develop and actively manage Public Awareness Programs. This document will also help to raise the quality of pipeline operators' Public Awareness Programs, establish consistency among such programs throughout the pipeline industry, and provide mechanisms for continuous improvement of the programs. The recommended practice has been developed specifically for pipelines operating in the United States, but may also have use in international settings.
		This RP identifies for the pipeline operator four specific stakeholder audiences and associated public outreach messages and communication methods to choose from in developing and managing a successful Public Awareness Program. It also provides information to assist operators in establishing specific plans for public awareness that can be evaluated and updated.
		This document provides guidance for the following pipeline operators: intrastate and interstate hazardous liquid pipelines; intrastate and interstate natural gas transmission pipelines; local distribution systems, and gathering systems. This guidance is intended for use by pipeline operators in developing and implementing Public Awareness Programs associated with the normal operation of existing pipelines.
		The guidance is not intended to focus on public awareness activities appropriate for new pipeline construction or for communications that occur immediately after a pipeline related emergency. Communication regarding construction of new pipelines is highly specific to the type of pipeline system, scope of the construction, and the community and state in which the project is located. Likewise, public communications in response to emergency situations are also highly specific to the emergency and location. This RP is also not intended to provide guidance to operators for communications about operator-specific performance measures that are addressed through other means of communication or regulatory reporting.
		The primary audience for this document is the pipeline operator for use in developing a Public Awareness Program for the following stakeholder audiences: the affected public—i.e., residents, and places of congregation (businesses, schools, etc.) along the pipeline and the associated right-of-way (ROW); local and state emergency response and planning agencies [i.e., State and County Emergency Management Agencies (EMA) and Local Emergency Planning Committees (LEPCs)]; local public officials and governing councils; and, excavators.

Standard No	Title	Description
API 1163	In-line inspection systems qualification standard edition	Published by the American Petroleum Institute (API), it covers the use of in-line inspection systems for onshore and offshore gas and hazardous liquid pipelines. It includes, but is not limited to, tethered or free-flowing systems for detecting metal loss, cracks, mechanical damage, pipeline geometries, and pipeline location or mapping, The document applies to both existing and developing technologies. This Standard is an umbrella document that provides performance-based requirements for in-line inspection systems, including procedures, personnel, equipment, and associated software.
API 1164	SCADA security	Published by the American Petroleum Institute, it provides guidance to the operators of oil and gas liquid pipeline systems for managing SCADA system integrity and security. The use of this document is not limited to pipelines regulated under Title 49 CFR 195.1, but should be viewed as a listing of best practices to be employed when reviewing and developing standards for a SCADA system. This document embodies the "API Security Guidelines for the Petroleum Industry." This guideline is specifically designed to provide the operators with a description of industry practices in SCADA Security, and to provide the framework needed to develop sound security practices within the operator's individual companies. It is important that operators understand system vulnerability and risks when reviewing the SCADA system for possible system improvements.
API RP 1165	Recommended practice for pipeline SCADA displays	Published by the American Petroleum Institute, this RP focuses on the design and implementation of displays used for the display, monitoring, and control of information on pipeline Supervisory Control and Data Acquisition Systems (SCADA). The primary purpose is to document industry practices that provide guidance to a pipeline company or operator who want to select a new SCADA system, or update or expand an existing SCADA system. This document assists pipeline companies and SCADA system developers in identifying items that are considered best practices when developing human machine interfaces (HMI). Design elements that are discussed include, but are not limited to, hardware, navigation, colors, fonts, symbols, data entry, and control / selection techniques.

Standard No	Title	Description
API RP 1166	Excavation monitoring and observation	Published by the American Petroleum Institute, it provides a consistently applied decision-making process for monitoring and observing of excavation and other activities on or near pipeline rights-of -way for "hazardous liquid" and "natural and other gas" transmission pipelines. (Note: One-call provisions and laws vary by state, and it is the operator's responsibility to be familiar with and comply with all applicable one-call laws.). This document's purpose is to protect the public, excavation employees, and the environment by preventing damage to pipeline assets from excavation activities.
API RP 1168	Pipeline control room management	Published by the American Petroleum Institute, it provides pipeline operators and pipeline controllers with guidance on industry best practices on control room management to consider when developing or enhancing practices and procedures. This document was written for operators with continuous and non-continuous operations, as applicable. It addresses four pipeline safety elements for hazardous liquid and natural gas pipelines in the transportation and distribution sectors: pipeline control room personnel roles, authorities and responsibilities; guidelines for shift turnover; pipeline control room fatigue management; and, pipeline control room management of change (MOC).
API RP 2200	Repairing crude oil, liquefied petroleum gas and product pipelines	Published by the American Petroleum Institute, it discusses guidelines to safe practices while repairing pipelines for crude oil, liquefied petroleum gas, and product service. Although it is recognized that the conditions of a particular job will necessitate an on-the-job approach, the observance of the suggestions in this document should improve the probability that repairs will be completed without accidents or injuries.
B31.1	Power piping	The B31.1 code prescribes minimum requirements for the design, materials, fabrication, erection, test, inspection, operation, and maintenance of piping systems typically found in electric power generating stations, industrial institutional plants, geothermal heating systems, and central and district heating and cooling systems. The code also covers boiler external piping for power boilers and high temperature, high pressure water boilers in which steam or vapor is generated at a pressure of more than 15 psig; and high temperature water is generated at pressures exceeding 160 psig and/or temperatures exceeding 250 degrees F.
B31.11	Slurry transportation piping systems:	The primary purpose of this Code is to establish requirements for design, construction, inspection, testing, operation, and maintenance of slurry transportation piping systems for protection of the general public and operating company personnel, as well as for reasonable protection of the piping system against vandalism and accidental damage by others, and reasonable protection of the environment. (5)

Standard No	Title	Description
B31.2	Fuel gas piping	Was part of the B31 ASME code for pressure piping. However, it was withdrawn and replaced by as a National Standard and replaced by ANSI/NFPA Z223.1 (5).
B31.3	Process pipeline	The B31.3 code contains requirements for piping typically found in petroleum refineries; chemical, pharmaceutical, textile, paper, semiconductor, and cryogenic plants; and related processing plants and terminals. This code prescribes requirements for materials and components, design, fabrication, assembly, erection, examination, inspection, and testing of piping. This Code applies to piping for all fluids including: (1) raw, intermediate, and finished chemicals; (2) petroleum products; (3) gas, steam, air and water; (4) fluidized solids; (5) refrigerants; and (6) cryogenic fluids. Also included is piping which interconnects pieces or stages within a packaged equipment assembly. (5).
B31.4	Pipeline transportation systems for liquid hydrocarbons and other liquids	This standard prescribes requirements for the design, materials, construction, assembly, inspection, and testing of piping transporting liquids between production facilities, tank farms, natural gas processing plants, refineries, pump stations, ammonia plants, terminals (marine, rail, and truck), and other delivery and receiving points. Piping consists of pipe, flanges, bolting, gaskets, valves, relief devices, fittings, and the pressure-containing parts of other piping components. It also includes hangers and supports, and other equipment items necessary to prevent overstressing the pressure-containing parts. It does not include support structures such as frames of buildings, stanchions, or foundations, or any equipment. Requirements for offshore pipelines are found in Chapter IX. Also included within the scope of this Code are • (a) primary and associated auxiliary liquid petroleum and liquid anhydrous ammonia piping at pipeline terminals (marine, rail, and truck), tank farms, pump stations, pressure reducing stations, and metering stations, including scraper traps, strainers, and prover loops • (b) storage and working tanks, including pipe-type storage fabricated from pipe and fittings, and piping interconnecting these facilities • (c) liquid petroleum and liquid anhydrous ammonia piping located on property which has been set aside for such piping within petroleum refinery, natural gasoline, gas processing, ammonia, and bulk plants • (d) those aspects of operation and maintenance of Liquid Pipeline Systems relating to the safety and protection of the general public, operating company personnel, environment, property, and the piping systems.(5)

Standard No	Title	Description
B31.5	Refrigeration piping and heat transfer components:	This Code contains requirements for the materials, design, fabrication, assembly, erection, test, and inspection of refrigerant, heat transfer components, and secondary coolant piping for temperatures as low as -320°F (-196°C), whether erected on the premises or factory assembled. Users are advised that other piping Code Sections may provide requirements for refrigeration piping in their respective jurisdictions. This Code does not apply to: (a) any self- contained or unit systems subject to the requirements of Underwriters Laboratories or other nationally recognized testing laboratory: (b) water piping; (c) piping designed for external or internal gage pressure not exceeding 15 psi (105 kPa) regardless of size; or (d) pressure vessels, compressors, or pumps, but does include all connecting refrigerant and secondary coolant piping starting at the first joint adjacent to such apparatus.
B31.8	Gas transmission and distribution piping systems	The B31.8 Code covers the design, fabrication, installation, inspection, testing, and other safety aspects of operation and maintenance of gas transmission and distribution piping systems, including gas pipelines, gas compressor stations, gas metering and regulation stations, gas mains, and service lines up to the outlet of the customer's meter set assembly. The scope of this Code includes gas transmission and gathering pipelines, including appurtenances, that are installed offshore for the purpose of transporting gas from production facilities to onshore locations; gas storage equipment of the closed pipe type that is fabricated or forged from pipe or fabricated from pipe and fittings; and gas storage lines.
B31.8S	Managing system integrity of gas pipelines	This Standard applies to on-shore pipeline systems constructed with ferrous materials and that transport gas. This includes all parts of physical facilities through which gas is transported, including pipe, valves, appurtenances attached to pipe, compressor units, metering stations, regulator stations, delivery stations, holders and fabricated assemblies. The principles and processes embodied in integrity management are applicable to all pipeline systems. This Standard is specifically designed to provide the operator with the information necessary to develop and implement an effective integrity management program utilizing proven industry practices and processes. The processes and approaches within this Standard are applicable to the entire pipeline system.

Standard No	Title	Description
B31G Manual	Manual for determining remaining strength of corroded pipelines	This document is intended solely for the purpose of providing guidance in the evaluation of metal loss in pressurized pipelines and piping systems. It is applicable to all pipelines and piping systems that are part of ASME B31 Code for Pressure Piping, i.e., ASME B31.4 Pipeline Transportation Systems for Liquid Hydrocarbons and Other Liquids; ASME B31.8, Gas Transmission and Distribution Piping Systems; ASME B31.11, Slurry Transportation Piping Systems; and ASME B31.12, Hydrogen Piping and Pipelines, Part PL.

References

1. Cartz, L. *Nondestructive Testing*. Materials Park : ASM International, 1995. All rights reserved. Used with permission.
2. US Department of Transportation. www.access.gpo.gov/nara/cfr/waisidx_00/49cfr192_00.html. (website)
3. Lawson, K. *Direct Assessment Methods for Pipeline Integrity Management*. Clarion Technical Publishers, 2007. Short course document.
4. UK Health and Safety Executive: *ALARP at a Glance*. UK Health and Safety Executive Web site. (website) [Cited: 09 19, 2010.] Contains public sector information published by the Health and Safety Executive and licensed under the Open Government Licence v1.0.
5. Bianchetti, R.E. (editor). *Peabody's Control of Pipeline Corrosion, 2nd Edition*. ©NACE International, 2001. All rights reserved. Reprinted with permission.
6. Bruce, B. and Alexander, C. *Pipeline Repair Methods/In-Service Welding*. Clarion Technical Publishers, 2011. Short course document.
7. Tiratsoo, J. (editor). *Pipeline Pigging & Integrity Technology, Third Ed.* Scientific Surveys Ltd & Clarion Technical Publishers 2003.
8. Hopkins, P. *Defect Assessment in Pipelines*. Penspen Group, 2011. Short course document.
9. Mohitpour, M., Golshan, H., and Murray, A. *Pipeline Design & Construction, A Practical Approach, Third Ed.* ASME Press, 2007.
10. Hopkins, P., King, R., and Miesner, T. *Onshore Pipeline Engineering*. Clarion Technical Publishers, 2011. Short course document.
11. Energy Resources Conservation Board. "Defining Sour Gas." ERCB (website). www.ercb.ca.
12. Jones, D. *Principles and Prevention of Corrosion, 2nd Edition*. Prentice-Hall, 1996.
13. Bull, D.E. and Williamson, G. *DOT Pipeline Safety Regulations*. Clarion Technical Publishers, 2011. Short course document.
14. Argent, C. (editor). *Macaw's Pipeline Defects*. Yellow Pencil Marketing Co. Ltd, 2003.
15. Desaulniers, R. and Marr, J. *Excavation Inspection & Applied NDE for ILI/DA Validation and Correlation*. Clarion Technical Publishers, 2011. Short course document.
16. *Specifications and Requirements for Intelligent Pig Inspection of Pipelines, Version 2009*, Pipeline Operators Forum (www.pipelineoperatorsforum.org).
17. Shell International Exploration and Production B.V. Specification and Requirements for Intelligent Pig Inspection of Pipelines. 1998.
18. King, R. *Microbiological Corrosion of Pipelines* Course Notes. Clarion Technical Publishers, 2011. Short course document.
19. Gelman, A., et al. *Bayesian Data Analysis*. Chapman & Hall, 1995.
20. Baniak, E. *API Spec 5L, 44th Edition: Specification for Line Pipe*. API (website). October 1, 2008. www.api.org/meetings/topics/pipeline/upload/Ed_Baniak_Technical.pdf.
21. Argent, C. (editor). *Macaw's Pipeline Defects*. Yellow Pencil Marketing Co. Ltd, 2003.

22. Hopkins, P. "Introduction to Basic Pipeline Engineering." In: Hopkins, P., King, R., and Miesner, T. *Onshore Pipeline Engineering* Clarion Technical Publishers, 2011. Short course document.

23. Barrett, C.R., Nix, W.D., and Tetelman, A.S. *The Principles of Engineering Materials, First Ed.* © 1973. Reprinted with permission of Pearson Education, Inc., Upper Saddle River, NJ.

24. Canadian Energy Pipeline Association (CEPA). *Stress Corrosion Cracking Recommended Practice, 2nd Edition*. Canadian Energy Pipeline Association, 2007.

25. Fessler, R. and Mackenzie, J. *Stress Corrosion Cracking in Pipelines*. Clarion Technical Publishers, 2011. Short course document.

26. Rizkalla, M. (editor). *Pipeline Geo-Environmental Design and Geohazard Management*. ASME Press, 2008.

27. Mohitpour, M., Szabo, J. and Van Hardeveld, T. *Pipeline Operation & Maintenance: A Practical Approach, Second Ed.* ASME Press, 2010.

28. American Petroleum Institute. 2009 Publications Programs and Services (website). [Cited: 09 03, 2009.] global.ihs.com/images/ENGL/api_catalog.pdf?currency_code=USD&customer_id=2125482C5B0A&shopping_cart_id=282438372E4A50384B5A5D28260A&rid=API1&mid=Q023&country_code=US&lang_code=ENGL&input_doc_number=&input_doc_title=.

29. ASME International. ASME standards of pressure piping. (website) [Cited: April 29, 2009.] www.engineeringtoolbox.com/asme-b31pressure-piping-d_39.html.

30. Romanoff, M. *Underground Corrsosion*. ©NACE International, 1989. All rights reserved. Reprinted with permission.

31. Muhlbauer, W.K. *Pipeline Risk Management Manual, Third Ed.* Elsevier, 2004.

32. Brown, E.T. (editor). *Analytical and Computational Methods in Engineering Rock Mechanics*. Allen & Unwin, 1987

33. Squires, G.L. *Practical Physics*. Cambridge University Press, 1985.

34. Zachs, S. *Introduction to Reliability Analysis*. Springer Verlag, 1992.

35. Nessim, M., Zimmerman, T. and Fuglem, M. *Guidelines for Reliability-Based Design and Assessment of Onshore Pipelines*.

36. Hogg, R., McKean, J. and Craig, A. *Introduction to Mathematical Statistics*. Macmillan, 1978.

37. Moaveni, S. *Finite Element Analysis: Theory and Application with ANSYS*. Prentice Hall, 2007.

38. National Energy Board. "Living and Working Near Pipelines: Landowner Guide 2005." National Energy Board. (website) 2005. [Cited: November 10, 2008.] www.neb.gc.ca/clf-nsi/rsftyndthnvrnmnt/sfty/rfrncmtrl/lvngwrkngnrpplnsgd-eng.pdf.

39. American Society for Non-Destructive Testing. "Learning: What Is NDT?" ASNT Website. (website) Unknown, November 12, 2008. [Cited: November 12, 2008.] www.asnt.org/ndt/primer1.htm.

40. Westwood, S. and Hopkins, P. Products and Services: Pipeline Inspection - Technical Papers: Pigging Defect Assessment Codes. BJ Pipeline & Process Services (now part of Baker Hughes Pipeline Management Group). Corporate Website (website). [Cited: 11 12, 2008.] www.bjservices.com/website/pps.nsf/WebPages/PISTechPapers/$file/VECTRA%20Pigging%20Defect%20Assessment%20Codes.pdf?OpenElement.

41. NACE International Store: Products by Industry: Pipeline/Underground Tanks:NACE Standards. NACE International Website (website). [Cited: 11 12, 2008.] web.nace.org/Departments/Store/product.aspx?id=917bf186-3502-4d99-b73c-425bda3defba.

42. Eldevik, F. "Safe Pipeline Transmission of CO2." *Pipeline & Gas Journal*. November 2008.

43. CSA. CSA Product Store. Z245.1 (website). [Cited: February 13, 2009.] www.csa-intl.org/onlinestore/GetCatalogItemDetails.asp?mat=2418620&Parent=419.

44. NACE International website: NACE Standards: Individual Standards and Packaged Sets: All Standards. [Cited: 02 February, 2009.] web.nace.org/Departments/store.

45. NACE International website: Reports. [Cited: February 20, 2009.]

46. NACE International website: About Us. [Cited: February 20, 2009.] www.nace.org/content.cfm?parentid=1005&currentID=1005.

47. Office of Pipeline Safety website: About Us. PHMSA. [Cited: February 13, 2009.] phmsa.dot.gov/pipeline/about.

48. Bianchetti, R.L. "Impressed Current Cathodic Protection, " in *Peabody's Control of Pipeline Corrosion, Second Ed.* ©NACE International, 2001. All rights reserved. Reprinted with permission.

49. Atkins, P. *Physical Chemistry*. W.H. Freeman and Co., 1998.

50. American National Standards Institute. Introduction. About ANSI (website). [Cited: April 29, 2009.] www.ansi.org/about_ansi/introduction/introduction.aspx?menuid=1.

51. Wilkins, D.J. "The Bathtub Curve and Product Failure Behavior (Part 1 of 2)." Reliability Hotwire (website). November 2002. [Cited: April 29, 2009.] www.weibull.com/hotwire/issue21/hottopics21.htm.

52. American Gas Association. AGA: About AGA. American Gas Association (website). [Cited: April 30, 2009.] www.aga.org/About/.

53. American Society for Nondestructive Testing. What Is ASNT. American Society for Nondestructive Testing. (website) [Cited: April 30, 2009.] www.asnt.org/asnt/asnt.htm.

54. Helmenstine, A. About.com: Chemistry. About.com (website). [Cited: May 1, 2009.] chemistry.about.com/od/chemistryglossary/.

55. Schlumberger. Oilfield Glossary (website). [Cited: May 3, 2009.] www.glossary.oilfield.slb.com/.

56. Chikezie, N. "Improving Valve Performance," *Pipeline and Gas Technology*. March 2009.

57. NACE International. *NACE Corrosion Engineer's Reference Book*. ©NACE International, 2002. All rights reserved. Reprinted with permission.

58. Stanczak, Marianne. «Biofouling: It›s Not Just Barnacles Anymore» (website). [Cited: May 26, 2009.] www.csa.com/discoveryguides/biofoul/overview.php.

59. Tullmin, M. Corrosion Club (website). 2008. [Cited: 09 04, 2009.] www.corrosion-club.com/noise.htm.

60. Kittel, C. *Introduction to Solid-State Physics*. John Wiley & Sons, 1976.

61. Oxford University. Chemical Safety Information - Glossary (website). 04 2008. [Cited: 09 04, 2009.] msds.chem.ox.ac.uk/glossary/GLOSSARY.html.

62. ASM International. About ASM (website). 2009. [Cited: 09 04, 2009.] asmcommunity.asminternational.org/portal/site/www/About/.

63. About.com. Composite Materials Glossary. About.com (website). About.com, 2009. [Cited: 09 08, 2009.] composite.about.com/od/glossaries/l/blglossary_c.htm.

64. Marion, J.B. *Classical Dynamics of Particles and Systems*. Academic Press, 1970.

65. Lewis, M. "Instrumentation," in *Peabody's Control of Pipeline Corrosion, Second Ed.* ©NACE International, 2001. All rights reserved. Reprinted with permission.

66. US Department of Transportation. PHMSA (website). [Cited: 09 08, 2009.] www.phmsa.dot.gov/.

67. US Department of Energy (website). www.energy.gov/.

68. US Environmental Protection Agency. EPA (website). [Cited: 09 09, 2009.] www.epa.gov/.

69. Matrix Group International, Inc. AOPL, The Organization. AOPL. (website) Matrix Group International, Inc, 2009. [Cited: 09 11, 2009.] www.aopl.org/aboutAOPL/.

70. American Water Works Association. *C105-05 Polyethylene Encasement for Ductile-Iron Pipe Systems*. American Water Works Association. (website). [Cited: 09 12, 2009.] www.awwa.org/index.cfm.

71. NDT Resource Center (website). www.ndt-ed.org/index_flash.htm.

72. Rarrett, C.R., Nix, W., and Tetelman, A.S. *The Principles of Engineering Materials*. Prentice-Hall, 1973.

73. Halliday, D. and Resnick, R. *Fundamentals of Physics*. John Wiley & Sons, 1974.

74. Unified Engineering Inc. www.unified-eng.com/scitech/hardness/hardness.html. Unified Engineering Inc. (website) 2008. www.unified-eng.com/scitech/hardness/hardness.html.

75. Revie, W.R. (editor). *Uhlig's Corrosion Handbook, Third Ed*. John Wiley & Sons, 2011.

76. Scanlan, R.J., Boothman, R.M., and Clarida, D.R. *Corrosion Monitoring Experience in the Refining Industry Using the FSM Technique*. ©NACE International, 2003. All rights reserved. Reprinted with permission.

77. Bianchetti, R.L. «Survey Methods and Evaluation Techniques,» in *Peabody's Control of Pipeline Corrosion, Second Ed*. ©NACE International, 2001. All rights reserved. Reprinted with permission.

78. Heeger, D. Signal Detection Theory (website). 2007. [Cited: 09 13, 2009.] www.cns.nyu.edu/~david/handouts/sdt/sdt.html.

79. Huxhold, W.E. and Levinsohn, A.G. *Managing Geographic Information Systems Projects*. Oxford University Press, 1995.

80. Van Vlack, L.H. *Elements of Materials Science and Engineering*. Addison-Wesley, 1989.

81. Davidson College. Web Physics (website). webphysics.davidson.edu/.

82. Canadian Energy Producers Association. CEPA. About CEPA (website). www.cepa.com/.

83. Canadian Standards Association. CSA. About CSA (website). www.csa.ca/.

84. GMI Robinetterie Industrielle (website). www.gmi-valves.com/.

85. NACE SP #35100. *In-line Nondestructive Inspection of Pipelines*. ©NACE International, 2000. All rights reserved. Reprinted with permission.

86. Beller, M., Holden, E., and Uzelac, N. «Cracks in Pipelines and How to Find Them," in *Pipeline Pigging & Integrity Technology, Third Ed.,* J. Tiratsoo (Editor), Scientific Surveys Ltd & Clarion Technical Publishers 2003.

87. Timashev, S.A. *Risk-Based Management of Pipeline Integrity Safety*, Clarion Technical Publishers, 2011. Short course document.

88. 90. Nicholson, P. "External Corrosion Direct Assessment" (website). cath-tech.com/papers/corcopnindia2004.pdf.m

89. Gray, J.F. "Developing a Maintenance-Pigging Program," in *Pipeline Pigging & Integrity Technology, Third Ed.,* J. Tiratsoo (Editor), Scientific Surveys Ltd & Clarion Technical Publishers 2003.

90. Cameron, G.R. "Pigging 'Unpiggable' Pipelines: A Guide for Maintenance Pigging and Preparation for Smart Pig Inspection," in *Pipeline Pigging & Integrity Technology, Third Ed.,* J. Tiratsoo (Editor), Scientific Surveys Ltd & Clarion Technical Publishers, 2003.

91. Kershaw, C. "Enhanced Cleaning Is Cost Effective," in *Pipeline Pigging & Integrity Technology, Third Ed.,* J. Tiratsoo (Editor), Scientific Surveys Ltd & Clarion Technical Publishers 2003.

92. Smart, J. "Pigging and Chemical Treatment of Pipelines," in *Pipeline Pigging & Integrity Technology, Third Ed.,* J. Tiratsoo (Editor), Scientific Surveys Ltd & Clarion Technical Publishers, 2003.

93. Brown, G. "Latest Design Techniques for Dual- and Multi-Diameter Pipeline Pigs," in *Pipeline Pigging & Integrity Technology, Third Ed.,* J. Tiratsoo (Editor), Scientific Surveys Ltd & Clarion Technical Publishers, 2003.

94. Vieth, P. "Assessment Criteria for ILI Metal-Loss Data: B31G and RStreng," in *Pipeline Pigging & Integrity Technology, Third Ed.,* J. Tiratsoo (Editor), Scientific Surveys Ltd & Clarion Technical Publishers, 2003.

95. Wilder, J.R. "Batching an Ultrasonic Pig in a Natural Gas Liquids Pipeline," in *Pipeline Pigging & Integrity Technology, Third Ed.,* J. Tiratsoo (Editor), Scientific Surveys Ltd & Clarion Technical Publishers, 2003.

96. van Agthoven, R. "Inspection of Unpiggable Pipelines: Experience, History, and the Future of Cable-Operated Ultrasonic Pigging," in *Pipeline Pigging & Integrity Technology, Third Ed.,* J. Tiratsoo (Editor), Scientific Surveys Ltd & Clarion Technical Publishers, 2003.

97. Levorsen, A.I. *Geology of Petroleum.* W. H. Freeman Co., 1967.

98. API 1163. *In-line Inspection Systems Qualification Standard.* American Petroleum Institute, 2005.

99. Bianchetti, R.L. and McKim, S. "Ground Bed Design," in *Peabody's Control of Pipeline Corrosion, Second Ed.* ©NACE International, 2001. All rights reserved. Reprinted with permission.

100. Bianchetti, R. «Cathodic Protection with Galvanic Anodes," in *Peabody's Control of Pipeline Corrosion, Second Ed.* ©NACE International, 2001. All rights reserved. Reprinted with permission.

101. Sloan, R. «Pipeline Coatings,» in *Peabody's Control of Pipeline Corrosion, Second Ed* ©NACE International, 2001. All rights reserved. Reprinted with permission.

102. McManus, J.J., Pennie, W.L., and Davies, A. "Hot Applied Coal Tar Coatings," *Industrial & Engineering Chemistry*: Volume 58, Issue 4 pp. 43-46, April 1966.

103. Guan, S.W. Corrosion Protection by Coatings for Water and Wastewater Pipelines. Madison Chemical Industries (website). www.madisonchemical.com/pdf_tech_papers/AUCSC2001_The_Corrosion_Protection.pdf

104. Harve Group. Steel Pipe Coating Plant (website). www.harvegroup.com/images/steel-pipe-coating-plant1.jpg.

105. BJ Process & Pipeline Services. Pipeline Inspection Brochure (website). www.bjservices.com/.

106. Bianchetti, R.L. "Construction Practices," in *Peabody's Control of Pipeline Corrosion, Second Ed.* ©NACE International, 2001. All rights reserved. Reprinted with permission.

107. Friedman, G.M. and Sanders, J.E. *Principles of Sedimentology.* John Wiley & Sons, 1978.

108. Clock Spring Company, LP. About. Clock Spring (website). www.clockspring.com/.

109. Grant, F.S. and West, G.F. *Interpretation Theory in Applied Geophysics.* McGraw-Hill, 1965.

110. Petrosleeve (website). www.petrosleeve.com/ps/specifications.html.

111. DNV Columbus. Fitness for Purpose (website). www.fitness4service.com/.

112. *API 570. Piping Inspection Code.* American Petroleum Institute, 2003.

113. Bannantine, J.A., Comer, J.J., and Handrock, J.L. *Fundamentals of Metal Fatigue Analysis.* Prentice-Hall, 1990.

114. Mosely, Mark. *DAMA/DMBOK Functional Framework, v. 3.02* ©2008. Used with permission.

115. Ritter, D.F. *Process Geomorphology.* Wm. C. Brown Company Publishers, 1978.

116. Brown, R.J., in *Subsea Pipeline Engineering,* by Palmer, A.C., King, R.A., et al. Clarion Technical Publishers, 2011. Short course document.

117. Gas Technology Institute. About GTI: GTI. GTI Website (website). [Cited: 06 22, 2010.] www.gastechnology.org/webroot/app/xn/xd.aspx?it=enweb&xd=AboutGTI.xml.

118. GMI Robinetterie Industrielle. Glossary (website). 2010. www.gmi-valves.com/glossaire.php.

119. Guided Wave Analysis LLC. "Ultrasonic Guided Wave Inspection" (website). gwanalysis.com/long-range-UT-inspection-range.html.

120. Enerisq International (website). enerisq.org/EnerisQ_FAQs_GWUT.pdf.

121. Resnick, D. and Halliday, R. *Fundamentals of Physics.* John Wiley & Sons, 1974.

122. Yi-Qun, L. and O'Handley, R. "An Innovative Passive Solid-State Magnetic Sensor." Sensors (website). 2000. archives.sensorsmag.com/articles/1000/52/index.htm.

123. Allegro. "Hall-Effect IC Application Guide" (website). Allegro MicroSystems Inc. www.allegromicro.com/en/Products/Design/hall-effect-sensor-ic-applications-guide/.

124. Bubenik, T., Moreno, P., Smith, G, and Williamson, G. *Pipeline Pigging & In-Line Inspection*. Clarion Technical Publishers, 2011. Short course document.

125. About.com. Metals. Amout.com (website). metals.about.com/.

126. Interstate Natural Gas Association of America. About. www.ingaa.org (website). [Cited: 08 20, 2009.] www.ingaa.org.

127. Smart, J. "Operational Safety in Pigging," in *Pipeline Pigging & Integrity Technology, Third Ed.*, J. Tiratsoo (Editor), Scientific Surveys Ltd & Clarion Technical Publishers, 2003.

128. CAPP. Statistical Handbook for Canada's Upstream Petroleum Industry. CAPP, 2010.

129. Accu Tech Consulting. AccuSafe December 1999 Newsletter. AccuSafe (website). Accu Tech Consulting, 12 1999. [Cited: 8, 20, 2009.] www.acusafe.com/Newsletter/Stories/1299News-PSM_Canada.htm.

130. Unified Engineering Inc. "Deciphering Weld Symbols." Unified Engineering Inc. (website). 2008. www.unified-eng.com/scitech/weld/weld.html.

131. Tiratsoo, J. (editor). *Pipeline Pigging & Integrity Technology, Third Ed.*, Scientific Surveys Ltd & Clarion Technical Publishers, 2003, p. 31.

132. Kreyszig, E. *Advanced Engineering Mathematics, Third Ed.* John Wiley and Sons, 1994

133. Pigging Products & Services Association. "About Pigs." Pigging Products & Services Association (website). www.ppsa-online.com/about-pigs.php.

134. Integrated Publishing. "Valves." Integrated Publishing (website). www.tpub.com/fireman/69.htm.

135. Mistras. "Mistras Services." Mistras Asset Protection Solutions (website). www.mistrasgroup.com/services/advancedndt/aut/gul.aspx.

136. GE Power Systems. "Crack Detection." GE Power Systems, Oil & Gas, PII Pipeline Solutions (website). www.geenergy.com/prod_serv/serv/pipeline/en/downloads/crack_detection.pdf.

137. Encyclopedia Britannica (website). www.britannica.com/.

138. University of Calgary. "What Is Geomatics Engineering?" University of Calgary Website (website). [Cited: 06 22, 2010.] www.geomatics.ucalgary.ca/about/whatis.

139. Deutsches Institut fur Normung (DIN). *Paints and Varnishes - Additional Terms and Definitions to DIN EN ISO 4618*. DIN, 2007. DIN EN ISO 4618.

140. Porter, P.C. "In-line Technology Ready to Handle New Pipeline Challenges," *Pipeline & Gas Journal*, October 2006.

141. PRCI. "About Us; Membership." PRCI Website. (website) [Cited: August 24, 2009.] prci.org/inde.php/site/membership; prci.org/inde.php/about.

142. PODS Association. "About Us" (website). [Cited: 09 12, 2010.] www.pods.org/about/.

143. Kruger, J. "Electrochemistry Encyclopedia" (website). Ernest B. Yeager Center for Electrochemical Sciences (YCES) and the Chemical Engineering Department , Case Western Reserve University. [Cited: 04 18, 2010.] electrochem.cwru.edu/encycl/art-c02-corrosion.htm.

144. The Society for Protective Coatings. "About Us" (website). [Cited: 03 29, 2010.] www.sspc.org/about/history.html.

145. Lillig, D.B., Newbury, B.D., and Altstadt, S.A. *The Second (2008) ISOPE Strain-Based Design Symposium - A Review.* The International Society of Offshore and Polar Engineers (ISOPE), 2009. TPC-775.

146. Zavala-Olivares, Geraldo, et al. "Sulfate-Reducing Bacteria Influence on the Cathodic Protection of Pipelines that Transport Hydrocarbons." NACE International, 2003. Paper #03087.

147. Bai, Q., Bai, Y. *Subsea Pipelines and Risers*. Elsevier, 2005.

148. DeMers, M.N. *Fundamentals of Geographic Information Systems*. John Wiley & Sons, 1997.

149. Parker, M.E. and Peattie, E.G. *Pipeline Corrosion and Cathodic Protection, Third Edition*. Elsevier, 1999.

150. Tiratsoo, J. (editor). *Pipeline Pigging & Integrity Technology, Third Ed.*, Scientific Surveys Ltd & Clarion Technical Publishers, 2003.

151. National Energy Board (website). October 2008. www.neb-one.gc.ca/clf-nsi/rcmmn/hm-eng.html.

152. International Pipeline Conference Foundation (website). 2007. [Cited: 08 20, 2009.] www.internationalpipelineconference.com/.

153. API 1160. *Managing Systems Integrity for Hazardous Liquid Pipelines*. American Petroleum Institute, 2001.

154. NACE RP0102. *In-Line Inspection of Pipelines*. ©NACE International, 2002. All rights reserved. Reprinted with permission.

155. Knowledge Publications (website). Knowledge Publications. knowledgepublications.com/.

156. NACE RP0169. ©NACE International, 2002. All rights reserved. Reprinted with permission.

157. ASTM International. ASTM D5162 - 08 (website). www.astm.org/Standards/D5162.htm.

158. ASME B31-8. *Gas Transmission and Distribution Piping Systems*. ASME, 2007.

159. ASTM G102. *Standard Practice for Calculation of Corrosion Rates and Related Information from Electrochemical Measurements*. American Society for Testing and Materials. ASTM G102.

160. NACE RP0102. *In-Line Inspection of Pipelines*. ©NACE International, 2002. All rights reserved. Reprinted with permission.

161. Gardiner, M.S. and Flemister, S.C. *The Principles of General Biology*. Macmillan, 1967.

162. Foreman, G. and Larsen, K.R. "Improved Magnetic Flux Leakage Technology Enhances Pipeline Inspection." *Materials Performance*, December 2008.

163. Mohitpour, M., Jenkins, A., and Nahas, G. "A Generalized Overview of Requirements for the Design, Construction, and Operation of New Pipelines of CO_2 Sequestration." *Journal of Pipeline Engineering*, pp. 237-251. December 2008.

164. Pfaffenberger, R.C. and Patterson, J.H. *Statistical Methods for Business and Economics*. Richard D. Irwin, 1977.

165. EC21. "PE Modified Asphalt Tape/Bitumen Tape. EC21 Global B2B Market Place (website). www.ec21.com/product-details/PE-Modified-Asphalt-Tape-Bitumen--3516720.html.

166. PHMSA. "About, Agency" (website). [Cited: March 15, 2010.] www.phmsa.dot.gov/about/agency.

167. NACE RP0502. *Pipeline External Corrosion Direct Assessment*. ©NACE International, 2002. All rights reserved. Reprinted with permission.

168. Bianchetti, R.L., (Editor). *Peabody's Control of Pipeline Corrosion, Second Ed.* ©NACE International, 2001. All rights reserved. Reprinted with permission.

169. Everitt, B.S. *The Cambridge Dictionary of Statistics, Third Edition*. Cambridge University Press, 2006.

170. Ugural, A.C. and Fenster, S.K. *Advanced Strength and Applied Elasticity*. Prentice-Hall, 1995.

171. Goerz, K., et al. "A Review of Methods for Confirming Integrity of Thermoplastic Liners - Field Experiences." NACE International, 2004. Paper #04073.

172. Lam, C.N.C., et al. "A New Approach To High-Performance Polyolefin Coatings." NACE International, 2007. Paper #07023.

173. Larock, B.E., Jeppson, R.W., and Watter, G.Z. *Hydraulics of Pipeline Systems*. CRC Press, 1999.

174. McAllister, E.W. (editor). *Pipeline Rules of Thumb Handbook, Seventh Edition*. Elsevier, 2009.

175. Mohr, W. *Report on Strain Based Design of Pipelines*. U.S. Department of the Interior, Minerals Management Service; and U.S. Department of Transportation, Research and Special Programs Administration, 2003. Project #45892GTH.

176. Moles, M., Dubé, N. and Ginzel, E. "Pipeline Girth Weld Inspection Using Ultrasonic Phased Array." Pan American Conference for Non-Destructive Testing, 2003, #T-050.

177. Thompson, A. and Taylor, B.N. Guide for the Use of the International System of Units (SI). Gaithersburg, MD : National Institute of Standards and Technology, 2008.

178. Unified Engineering. www.unified-eng.com/scitech/hardness/hardness.html. www.unified-eng.com/scitech/weld/weld.html.

179. Avis, W., et al. *Gage Canadian Dictionary*. Gage Educational Publishing Company, 1983.

180. Kedge, C., Personal communication. ©2010 C. Kedge. Used with permission.

181. Gibson, M., de Vries, G., *World Pipelines*, October 2005. ©2005 Palladian Publications. Used with permission.

182. PHMSA. "Pipeline Regulations" (website). [Cited: August, 2010.] www.phmsa.dot.gov/pipeline/regs.

183. National Energy Board. "Pipeline Abandonment – A Discussion Paper on Technical & Environmental Issues" (website) 2005. [Cited: August 2008.] www.neb.gc.ca/clf-nsi/rsftyndthnvrnmnt/sfty/rfrncmtrl/pplnbndnmnttchnclnvrnmntl-eng.html.

184. NACE Online Store, Products by Industry, Pipeline/Underground Tanks. NACE Website. (website). [Cited: August 2008.] web.nace.org/Departments/Store/Department.aspx?id=61.

185. Nessim, M., and Zhou, W., *Guidelines for Reliability Based Design and Assessment of Onshore Natural Gas Pipelines*, GRI Report #GRI-04/0229, 2005.

186. U.S. Army Corps of Engineers. Public Works Technical Bulletin 420-49-37: *Cathodic Protection Anode Selection*. July 2001.

187. Gelman, A., Carlin, J., Stern, H. and Rubin, D., *Bayesian Data Analysis*, Chapman & Hall, 1995.

188. British Stainless Steel Association Website. "Calculation of pitting resistance equivalent numbers (PREN)" (website). [Cited: July 2011] www.bssa.org.uk/topics.php?article=111

189. The Society for Protective Coatings. "About Us: What is SSPC" (website). [Cited: July 2011.] www.sspc.org/about-us/about-us-what-is-sspc/

190. International Standards Organization (website). "About Us" and "ISO in figures." [Cited: July 2011.] www.iso.org/iso/about.htm www.iso.org/iso/about/iso_in_figures/iso_in_figures_1.htm

191. The Australian Pipeline Industry Association (website). [Cited: July 2011.] www.apia.net.au//

192. American Society of Mechanical Engineers (website). "About Us." [Cited: July 2011.] www.asme.org/about-asme

193. European Pipeline Research Group (website). "About Ourselves." [Cited: July 2011.] www.eprg.net/aboutgroup/about.html

194. Health and Safety Executive (UK) (website). "About HSE." [Cited: July 2011.] www.hse.gov.uk/aboutus/index.htm

195. King. R., in Hopkins, P., King, R., and Miesner, T. *The Onshore Pipeline Engineering Course*. Clarion Technical Publishers, 2011. Short course document.

196. Muhlbauer, W.K., *Pipeline Risk Management Course*, Clarion Technical Publishers, 2011. Short course document.

197. National Welding Inspection School. *Welding Inspector's Handbook*, 2004.

198. ASTM E 436 – 91 (Reapproved 1997*). Standard Test Method for Drop-Weight Tear Tests of Ferritic Steels.*

199. Federal Highway Administration Report RD-01-156, *Cost of Corrosion in the USA*, 2001. Prepared by CC Technologies (now DNV Columbus). Downloadable at www. corrosioncost.com.

200. Crane Company. Reprinted with permission. All Rights Reserved.

201. Koppl Pipeline Services. Hot Tapping," (website) www.koppl.com/pdfs/Hottap.pdf Reprinted with permission.